AF734800

Power Quality - New Insights

Edited by Muhammad Mokhzaini Azizan

Published in London, United Kingdom

Power Quality - New Insights
http://dx.doi.org/10.5772/intechopen.110259
Edited by Muhammad Mokhzaini Azizan

Contributors
Ali Samari, Arief Goeritno, Bhavya Kanchanapalli, Irham Muhammad Azama, Juan Andrés Martín García, Makawi Diab Hraiz, Mohammad Ali Rahimi, Mohammad Mehdi Amiri, Muhammad Mokhzaini Azizan, Saleh Aghajan-Eshkevari, Sujatha Banka, Tapip Hendrawan

© The Editor(s) and the Author(s) 2024

The rights of the editor(s) and the author(s) have been asserted in accordance with the Copyright, Designs and Patents Act 1988. All rights to the book as a whole are reserved by INTECHOPEN LIMITED. The book as a whole (compilation) cannot be reproduced, distributed or used for commercial or non-commercial purposes without INTECHOPEN LIMITED's written permission. Enquiries concerning the use of the book should be directed to INTECHOPEN LIMITED rights and permissions department (permissions@intechopen.com).

Violations are liable to prosecution under the governing Copyright Law.

CC BY

Individual chapters of this publication are distributed under the terms of the Creative Commons Attribution 3.0 Unported License which permits commercial use, distribution and reproduction of the individual chapters, provided the original author(s) and source publication are appropriately acknowledged. If so indicated, certain images may not be included under the Creative Commons license. In such cases users will need to obtain permission from the license holder to reproduce the material. More details and guidelines concerning content reuse and adaptation can be found at http://www.intechopen.com/copyright-policy.html.

Notice

Statements and opinions expressed in the chapters are these of the individual contributors and not necessarily those of the editors or publisher. No responsibility is accepted for the accuracy of information contained in the published chapters. The publisher assumes no responsibility for any damage or injury to persons or property arising out of the use of any materials, instructions, methods or ideas contained in the book.

First published in London, United Kingdom, 2024 by IntechOpen
IntechOpen is the global imprint of INTECHOPEN LIMITED, registered in England and Wales, registration number: 11086078, 167-169 Great Portland Street, London, W1W 5PF, United Kingdom

British Library Cataloguing-in-Publication Data
A catalogue record for this book is available from the British Library

Additional hard and PDF copies can be obtained from orders@intechopen.com

Power Quality - New Insights
Edited by Muhammad Mokhzaini Azizan
p. cm.
Print ISBN 978-0-85466-182-4
Online ISBN 978-0-85466-181-7
eBook (PDF) ISBN 978-0-85466-183-1

We are IntechOpen, the world's leading publisher of Open Access books

Built by scientists, for scientists

7,000+
Open access books available

187,000+
International authors and editors

205M+
Downloads

156
Countries delivered to

Our authors are among the
Top 1%
most cited scientists

12.2%
Contributors from top 500 universities

WEB OF SCIENCE™

Selection of our books indexed in the Book Citation Index in Web of Science™ Core Collection (BKCI)

Interested in publishing with us?
Contact book.department@intechopen.com

Numbers displayed above are based on latest data collected.
For more information visit www.intechopen.com

For EU product safety concerns:
IN TECH d.o.o., Prolaz Marije Krucifikse Kozulić 3, 51000 Rijeka, Croatia,
info@intechopen.com or visit our website at intechopen.com.

Meet the editor

Muhammad Mokhzaini Azizan is an Associate Professor at Universiti Sains Islam Malaysia, and his work revolves around Engineering fields, majoring in Power and Energy, Renewable Technologies, Artificial Intelligence, Wireless Power Transfer and Security, and Image Processing, among others. His doctoral thesis focused on Power Transformers and Power Quality and has since produced a wide range of solutions to power quality-related problem statements. Due to rapid advancements in intelligence in systems, he embraced Artificial Intelligence, Image Processing, etc for a greater impact on the communities. He continues to produce translational research for wider communities, for example through the use of AI and ML, the behavior of a herd of elephants can be tracked, and innovations such as dryers and machines are benefitting small and medium enterprises in Malaysia. He was the head of an innovation company that provided intelligent solutions, intellectuals, etc for various businesses, through strategic ventures between academic and industry players. He is also a Malaysian Professional Engineer and Professional Technologist since 2020. He was appointed Deputy Dean (Research and Innovation) in 2023 and is currently working to facilitate research and innovation profile at the Faculty of Engineering and Built Environment.

Contents

Preface

This Edited Volume is a collection of reviewed and relevant research chapters focusing on developments within the field of power quality. The book includes scholarly contributions by various authors and is edited by a group of engineering experts. Each contribution comes as a separate chapter complete in itself but directly related to the book's topics and objectives.

The book includes the following chapters:

1. Introductory Chapter: Power Quality

2. Application-Based Simulations for Diagnosing the Power Quality When Changes in Load's Power on Each Bus as an Analogy of the Steady-State Stability Phenomena

3. Power Quality Improvement in Distribution System

4. Harmonics Emissions and Power System Grid Resilience in Electric Vehicle Charging

5. A Review on Utilization of Electric Vehicles for Mitigating the Power Quality Issues in Power Systems

IntechOpen Team

Chapter 1

Introductory Chapter: Power Quality

Muhammad Mokhzaini Azizan

1. Introduction

Traditionally, power quality has been a critical aspect of electrical engineering and energy management, focusing on the characteristics and reliability of the electrical power supply. The term encompasses a range of factors that collectively define the quality of electrical power as it is delivered from the generation source to end users. Over the years, the understanding and importance of power quality have evolved, reflecting changes in technology, industrial processes, and the increasing sensitivity of modern electronic equipment to variations in the electrical supply.

In the early stages of electrification, the emphasis was primarily on generating and distributing power over long distances. The key concern was to provide a steady and reliable supply of electrical energy to support the growth of industries and meet the increasing demand for electricity in homes and businesses. During this period, the concept of power quality was more centered on maintaining a stable voltage and frequency within acceptable limits. As technology advanced and the types of electrical loads diversified, power quality concerns expanded beyond just maintaining voltage and frequency. The proliferation of electronic devices and the integration of sophisticated control systems in industrial processes introduced new challenges. These challenges included issues related to harmonic distortions, voltage sags, transients, and other phenomena that could negatively impact the performance of sensitive equipment [1, 2]. The advent of information technology further elevated the importance of power quality. Computers, servers, and communication equipment became integral components of various industries, and these devices proved to be highly sensitive to fluctuations in the power supply [3, 4]. Even minor disturbances could lead to data corruption, system failures, and financial losses. This shift in the nature of loads necessitated a more comprehensive approach to power quality management.

Over time, advancements in technology have offered a myriad of solutions to address power quality issues. From the development of sophisticated power conditioning devices to the implementation of smart grid technologies, the field has seen continuous innovation. These technological solutions aim to enhance the resilience of power systems, mitigate disturbances, and ensure a more stable and reliable supply of electricity [2]. In the contemporary landscape, power quality faces new challenges, including the integration of renewable energy sources, the electrification of transportation, and the increasing complexity of interconnected systems [5, 6]. The future of power quality management lies in a holistic and adaptive approach, leveraging cutting-edge technologies like artificial intelligence for predictive maintenance, smart grids for dynamic load management, and energy storage for grid stabilization [7].

2. Traditional challenges

Over the years, traditional challenges related to power quality have persisted, requiring a comprehensive understanding of their causes, impacts, and effective solutions. They include a set of issues that stem from the dynamic nature of electrical systems, technological advancements, and the evolving needs of modern industries [8]. Sagging and swelling of grid voltage, which is also known as voltage variations, can occur due to various reasons such as grid faults, sudden load changes, or system disturbances. These variations can adversely affect sensitive equipment, leading to malfunctions or disruptions. Grid voltage swelling and overvoltage are commonly caused by changes in grid loads, especially large loads, and switching of power line. By conventional definition, a swell is defined as an increase in RMS voltage or current of between 1.1 and 1.8 pu at a duration of power quality ranging within a system frequency of 30 to 60 seconds event. Should the voltage reaches peak values, electrical devices are prone to damages and, in the worst case, permanent malfunction. By definition, an overvoltage is an operating condition where grid voltage is detected to increase and exceed its design limitation. The events that describe this phenomenon are transient, spikes, and so forth, and they can occur in either very short or long durations, depending on their severity of the condition. For example, switching overvoltage usually causes voltage oscillations with high value and high-frequency characteristics and often in short duration. In the cases of neutral conductor breakdown, overvoltage is produced in high voltage and results in long duration disruption, causing permanent damages to the grid. In both swell and overvoltage or sag conditions, the variations of voltage levels have been proven to adversely affect devices and equipment of power grid, whether at supply or load ends, or even at transmission lines. In most severe cases, power lines dielectric properties are compromised and could result in major breakdowns with energy being cut off, machines and motors are not working due to low supply, premature failures of systems' capacitor, control circuits, and so on could disrupt the grid operations, and so forth. In solving the conditions of voltage varieties (whether they were by design or results of operational issues), voltage regulation exercise is very important. An AVR is an example of devices commonly used in regulating voltage at the grid as it used in generators to stabilize and maintain voltage within acceptable limit. Devices such as surge protector and transient suppressor are installed in many locations at grid system in order to divert overly high voltage to keep grid equipment safe and operational. Voltage regulation can also be achieved by using devices such as tap changers, reactors, static VAR compensators, and so on.

Ever since inventors found alternating current (AC) electrical systems, Harmonics has been a concerning matter, impacting the grid and the electrical devices such as power cable, power transformer, protection relays, and so forth. Harmonics in power system are defined as sinusoidal components of a waveform that have frequencies that are integer multiples of the fundamental frequency of the system. Typically, in most power system, the fundamental frequency is either 50 or 60 Hz. Harmonics are additional frequencies that can be present due to types of loads that are nonlinear in nature. During the early days of AC system, harmonics existed due to the usage of arc furnaces and lightings due to industrialization. As the time goes, the types of loads evolved with more electronics are used in many systems be it at residential, offices, or commercial sites [6, 9]. The impacts of harmonics become more noticeable and significant due to these nonlinear types of loads [6, 8, 10]. In recent times where automations are taking central roles in the world, these nonlinear loads have been used even more such as rectifiers in electric vehicle (EV) applications, inverters in cooling systems, and variable drive systems such as VFD and VSD in machines and motor operations. Industrial processes, commercial operations, and quality of life at homes are undoubtedly improved and efficient; however, these devices draw current in a nonsinusoidal manner, contributing to greater harmonics distortion in power system. The effects of harmonics are profound as they exist in both voltage and current, which result in distortions in voltage or current. This can lead to increased losses in power grid, affecting performance of electronic-based equipment especially. With increases in loss, for example, Eddy current loss, equipment tend to operate with higher temperature than they were ever designed for in the first place. This overheating operation happens at power transformers, capacitors, and power cables and for a prolonged time without interventions; breakdowns of insulation may occur and result in permanent malfunction and breakdown of services [4, 10–12]. Mitigations on harmonics issues are well-defined over the years, with measures such as installation of harmonic filters and improved design of power transformer, sizing of cables, scheduling of load, and so forth being used. Passive and active filters are often designed to either minimize or simply filter out altogether the harmonics components in the system [4, 9]. By doing this, the nonsinusoidal waveforms are returned to the initial sinusoidal shape or reduced to specific desired threshold values, to be used as input to other devices without the risk of harmonics effects. To control the impact of harmonics to transformers, the design and sizing of the unit are configured to be able to withstand harmonic components. Special winding configuration, heat absorption or heat rejection mechanism, and improved insulations keep transformers operating without the risk of overheating and minimize the impact of harmonics. Several other measures include the scheduling of loads where loads that contribute to harmonics are distributed systematically throughout operational hours to reduce the concentration of harmonics component in the power system. By applying this method, losses due to harmonics are kept at a minimum and at an acceptable threshold value; hence, the risk of thermal overloading is reduced significantly.

3. Challenges in modern times

The advancements in science, technology, and engineering have no doubt contributed immensely to the world. Innovation and inventions through these fields have been the driving force behind solutions to address challenges that were never here before. Globally, discovery of knowledge allows human and societies to gain access to energy, sanitation, water, education, and so forth, which are crucial to transform the lives of billions and propel humanity forward in a sustainable nature.

A sustainable world includes the integration of renewable energy sources in the power system grid. Sources generating power through solar photovoltaic, wind power, and so on produce power into the conventional sourced power system grid, and the numbers are growing in large. These processes are complex and require the use of power electronics, which are loads that are nonlinear in nature. The electronics are used for converting variable outputs of renewable energy sources into a power system for consumption. Converters, rectifiers, and so on are producing distortions in power system current and voltage waveforms, thus causing harmonics to exist and impacting other devices. Renewable energy sources offer numerous benefits to sustainability in life and environment, but their widespread integration into power system grid produces great challenges in the context of power quality. Rapid switching of power electronic devices within inverters, for example, results in distortions and injection of nonsinusoidal current into a power system grid. Of course, the advantages of these applications to the world and people are far more than their advantages. Thus, it is essential for the systems to be designed and understood to ensure stable integration of renewable energy into a power system grid.

Modernization also involves massive use of an integrated system that can seamlessly provide high quality of life in a sustainable way. Transportation sector, which is among the biggest contributors of emissions in the world, opened opportunities for electric vehicles some time ago [13]. Now, the adaption of EV is at the highest peak, where the global EV outlook points to a 35% year-on-year increase with respect to new purchases in 2023 alone [14]. This also means that a wider charging network is being implemented, which, coupled with limit of traveling range and performance of the batteries, poses greater threat to power system grid in general [15]. Demand in electricity, especially during peak charging times, is rising exponentially at a rate never seen before. This means power system authorities and grid owners must manage this massive additional load, probably to allow more generation at the critical duration, and upgrade their transmission lines, transformers, and other infrastructures to cater more power and at the same time operate efficiently and profitably. Proper management of EV charging, perhaps in distributed evenly way can help reducing power system stress, to ensure stability and reliability [16, 17]. EV charging devices are becoming a major source of harmonics in transportation, due to power electronic converters used to convert AC to DC for charging the batteries. It also being extensively used to convert DC to AC, should the charging points utilize solar or wind or any other renewable sources. These converters generate harmonics current during the switching components during their operation. In some studies, there have been suggestions that to balance the need of EV and the negative impact of EV toward power quality, both EV owner and power system authorities/owner habits on this subject need to be adjusted accordingly. For example, charging EV battery with higher SOC could cause less distortion in a power system grid and the vehicle battery, locations of EV charging stations should not be in proximity to power transformers, the numbers of EV charging points are also to be limited to certain extent, and so forth.

4. Power quality – new insight

Researchers and industries have been exploring producing advanced inverter operational methodologies and strategies to control the impact of harmonics. Among others is the development of actively harmonic-blocking inverters, which are strategically distributed along a power system grid for mitigation processes [9, 18]. Monitoring power system parameters such as voltage and current state, power factor, transient conditions, and so on is done continuously to help identify potential issues way before their occurrences for implementing proactive mitigation measures. These measures are already in place nowadays with electrical devices such as relays and transformers being equipped with modern power quality analyzers with advanced capabilities for harmonic analysis, including artificial intelligence (AI) [18–22]. These are also known as advanced metering infrastructure (AMI), where deployments of smart meters communicate in two ways—collecting information and transferring it to a centralized center and also receiving information from the center for them to execute. AMI is one part of a wider surveillance capacity of a wide area monitoring system (WAMS), which is made of a strategic and synchronized system to view and provide action about dynamics of power system. WAMS provides comprehensive monitoring capability to facilitate greater active maintenance work. AI technologies are increasingly positioned in the field of power quality to enhance the monitoring, analysis, and management of a power system grid. Several contributions of AI into solutions of power quality include data analysis and pattern recognition, fault detection and diagnostics, predictive maintenance, power quality event classification and severity assessment, anomaly detection, and so on [18, 23, 24]. AI algorithms can analyze large datasets—including types of data and number of data—measurements, and pattern and trends are identified to recognize the potential of issues and produce predictions of the likelihood of power-quality events. AI can not only provide future event prediction; it can also be used for real-time fault diagnostics, where signals from sensor and instrument are measured and compared and provide insights of the causes [21, 22, 25]. In short, AI gives an edge of spotting and identifying the anomalies of power system operation parameters, quickly and accurately, round the clock and independently. Prompt investigation and resolution can be carried out, hence ensuring a more resilient and adaptive action. All these characteristics of AI operation give power system grid managers a great capacity, where the system can be made reliable and ready of any eventualities through sophisticated and automated solutions [21, 22, 25].

Author details

Muhammad Mokhzaini Azizan
Universiti Sains Islam Malaysia, Malaysia

*Address all correspondence to: mokhzainiazizan@usim.edu.my

IntechOpen

© 2024 The Author(s). Licensee IntechOpen. This chapter is distributed under the terms of the Creative Commons Attribution License (http://creativecommons.org/licenses/by/3.0), which permits unrestricted use, distribution, and reproduction in any medium, provided the original work is properly cited.

References

[1] Gbadamosi SL, Nwulu NI, Siano P. Harmonics Constrained Approach to Composite Power System Expansion Planning with Large-Scale Renewable Energy Sources. Energies. 2022;**15**:4070

[2] Eroğlu H, Cuce E, Cuce PM, Gul F, Iskenderoğlu A. Harmonic problems in renewable and sustainable energy systems: A comprehensive review. Sustainable Energy Technologies and Assessments. 2021;**48**:101566

[3] Liu Z, Sun Y, Xing C, Liu J, He Y, Zhou Y, et al. Artificial intelligence powered large-scale renewable integrations in multi-energy systems for carbon neutrality transition: Challenges and future perspectives. Energy and AI. 2022;**10**:100195

[4] Pathik BB, Islam S, Akram NM, Zaman MS, Al-Nahian MA. Total harmonic distortion reduction in a power grid: A UPQC based approach. In: 2023 3rd International Conference on Robotics, Electrical and Signal Processing Techniques (ICREST), Dhaka, Bangladesh. Piscataway, New Jersey, United States: Institute of Electrical and Electronics Engineers; 2023. pp. 264-268. DOI: 10.1109/ICREST57604.2023.10070091

[5] Shafiullah GM, Oo AMT. Analysis of harmonics with renewable energy integration into the distribution network. In: 2015 IEEE Innovative Smart Grid Technologies - Asia (ISGT ASIA), Bangkok, Thailand. Piscataway, New Jersey, United States: Institute of Electrical and Electronics Engineers; 2015. pp. 1-6. DOI: 10.1109/ISGT-Asia.2015.7387191

[6] Tümay M, Demirdelen T, Bal S, Kayaalp Rİ, Doğru B, Aksoy M. A review of magnetically controlled shunt reactor for power quality improvement with renewable energy applications. Renewable and Sustainable Energy Reviews. 2017;**77**:215-228

[7] Eltawil MA, Zhao Z. Grid-connected photovoltaic power systems: Technical and potential problems – A review. International Journal of Renewable and Sustainable Energy Review. 2010;**14**:112-129

[8] Cazacu E, Ioana G, Stănculescu M, Petrescu L. Active power filter solution to the harmonic issues caused by a nonlinear industrial load. In: 2023 10th International Conference on Modern Power Systems (MPS), Cluj-Napoca, Romania. Piscataway, New Jersey, United States: Institute of Electrical and Electronics Engineers; 2023. pp. 01-06. DOI: 10.1109/MPS58874.2023.10187576

[9] Schwanz D, Bagheri A, Bollen M, Larsson A. Active harmonic filters: Control techniques review. In: 2016 17th International Conference on Harmonics and Quality of Power (ICHQP), Belo Horizonte, Brazil. Piscataway, New Jersey, United States: Institute of Electrical and Electronics Engineers; 2016. pp. 36-41. DOI: 10.1109/ICHQP.2016.7783423

[10] Bagheri P, Xu W, Shaloudegi K. New indices to evaluate the impact of harmonic currents on power transformers. In: 2018 18th International Conference on Harmonics and Quality of Power (ICHQP), Ljubljana, Slovenia. Piscataway, New Jersey, United States: Institute of Electrical and Electronics Engineers; 2018. pp. 1-6. DOI: 10.1109/ICHQP.2018.8378885

[11] Sunbul A, Alduraibi A, Zare F. Harmonics Mitigation Filter for

High-Power Applications. IEEE Access. 2022;**10**:111406-111418

[12] Khazaee A, Chenarani H, Delavaripour H, Afshar R. Harmonic evaluation of distribution transformers based on real-time data of smart meters. In: 2023 27th International Electrical Power Distribution Networks Conference (EPDC), Mashhad, Iran, Islamic Republic of Iran. Piscataway, New Jersey, United States: Institute of Electrical and Electronics Engineers; 2023. pp. 90-93. DOI: 10.1109/EPDC59105.2023.10218808

[13] Alanazi F. Electric Vehicles: Benefits, challenges, and potential solutions for widespread adaptation. Applied Sciences. 2023;**13**:6016

[14] IEA. Global EV Outlook 2023. Paris: IEA; 2023. Available from: https://www.iea.org/reports/global-ev-outlook-2023

[15] Rusinaru D, Mircea PM, Buzatu GC. Power quality impact of electric buses' charging. In: 2023 Power Quality and Electromagnetic Compatibility at Low Frequency (PQEMC-LF), Craiova, Romania. Piscataway, New Jersey, United States: Institute of Electrical and Electronics Engineers; 2023. pp. 69-72. DOI: 10.1109/PQEMC-LF58184.2023.10211420

[16] Jadhav AB, Patil MS. A research based review on power quality improvement using evs with reference to smart grids. In: 2023 IEEE 5th International Conference on Cybernetics, Cognition and Machine Learning Applications (ICCCMLA), Hamburg, Germany. Piscataway, New Jersey, United States: Institute of Electrical and Electronics Engineers; 2023. pp. 237-241. DOI: 10.1109/ICCCMLA58983.2023.10346891

[17] Xu Y, Xu Y, Chen Z, Peng F, Beshir M. Harmonic analysis of electric vehicle loadings on distribution system. In: 2014 IEEE International Conference on Control Science and Systems Engineering, Yantai, China. Piscataway, New Jersey, United States: Institute of Electrical and Electronics Engineers; 2014. pp. 145-150. DOI: 10.1109/CCSSE.2014.7224526

[18] Dinesh M, Uthirasamy R, Suba S, Parathraju P. Improvement in power quality of a power systems using AI controller based filter. In: 2023 5th International Conference on Electrical, Computer and Communication Technologies (ICECCT), Erode, India. Piscataway, New Jersey, United States: Institute of Electrical and Electronics Engineers; 2023. pp. 1-6. DOI: 10.1109/ICECCT56650.2023.10179672

[19] Pandey U, Pathak A, Kumar A, Mondal S. Applications of artificial intelligence in power system operation, control and planning: a review. Clean Energy. 2023;**7**(6):1199-1218

[20] Oliveira d RA, Bollen MHJ. Deep learning for power quality. Electric Power Systems Research. 2023;**214**(Part A):108887

[21] Kuppusamy R, Nikolovski S, Teekaraman Y. Review of Machine Learning Techniques for Power Quality Performance Evaluation in Grid-Connected Systems. Sustainability. 2023;**15**(20):15055

[22] Thakur AK, Singh S, Singh SP. Machine learning based detection and classification of power quality disturbances in smart grid system. In: 2022 IEEE Global Conference on Computing, Power and Communication Technologies (GlobConPT), New Delhi, India. Piscataway, New Jersey, United States: Institute of Electrical and Electronics Engineers; 2022. pp. 1-6. DOI: 10.1109/GlobConPT57482.2022.9938352

[23] Abdubannaev J, Sun Y, Xin A, Jan MU, Makhamadjanova N, Rakhimov S. Enhancing power system resilience - a review. In: 2019 IEEE 3rd Conference on Energy Internet and Energy System Integration (EI2), Changsha, China. Piscataway, New Jersey, United States: Institute of Electrical and Electronics Engineers; 2019. pp. 2350-2354. DOI: 10.1109/EI247390.2019.9062002

[24] Iturrino Garcia CA, Bindi M, Corti F, Luchetta A, Grasso F, Paolucci L, et al. Power Quality Analysis Based on Machine Learning Methods for Low-Voltage Electrical Distribution Lines. Energies. 2023;**16**(9):3627

[25] Valencia-Duque AF, Alvarez Meza AM, Orizco-Gutierrez AA. Automatic identification of power quality events using machine learning approach. Scientia Et Technica. 2019;**24**:183-189

Chapter 2

Application-Based Simulations for Diagnosing the Power Quality When Changes in Load's Power on Each Bus as an Analogy of the Steady-State Stability Phenomena

Arief Goeritno, Tapip Hendrawan and Irham Muhammad Azama

Abstract

This chapter describes the system's ability to recover frequency synchronisation across all networks after a gradual change in electric power is determined by simulating the phenomenon of steady-state stability. The simulation is done by choosing and figuring out an analogy of the structure of an electric power network with four connected buses and assuming values for a number of parameters. The objectives for this diagnostic are set on a pair of primary levels, namely, to create the three syntax structures of the programme based on MATLAB and to simulate a phenomenon of steady-state stability based on the programme. The results of the implementation of steady-state stability simulation have been obtained in the form of phenomena such as a reduction of power on bus #1, an addition of power on bus #2, an addition of power on bus #3, and a reduction of power on bus #4. Based on the simulation results, it can be concluded that the MATLAB-based programme is used for the diagnosis of steady-state stability.

Keywords: application-based simulation, diagnosing the power quality, changes in load's power on each bus, steady-state stability phenomena, network structure with four buses

1. Introduction

MATLAB-based application is an interactive system and an application with a programming language. The basic data element is a matrix that does not need to declare the size or type of data, so many calculation problems can be solved in a short time, and calculations can be written into the FORTRAN or C language [1, 2]. Workarounds for complex calculation cases, such as power flow studies, make it difficult to change one value or more variables or commands for reevaluation. How to solve complex cases easily and efficiently to problems in the MATLAB command windows, done through providing input into a text file commonly called script files or M-files [1]. For the creation of an M-file with the selection of a new M-file from the File Menu, a text editor window is displayed [3].

The term power quality in the IEEE Standard Dictionary of Electrical and Electronics Terms is defined as “the concept of powering and grounding sensitive electronic equipment in a manner that is suitable to the operation of that equipment” [4]. The quality of electrical power is a problem in the form of voltage deviations, frequencies, or phase angles that result in failures or operating errors in electrical equipment on the consumer side. The diagnosis of electrical power quality is an effort to determine the type of deviation through the examination of various forms of phenomena that occur. The quality of electrical power is characterised by a voltage waveform in accordance with established guidelines. Poor quality of electrical power is a cause of loss, both on the customer side and the electric power generation side. Various factors can cause poor power quality, which can include voltage, current, frequency, and load factors [5].

Stability in electrical power systems (EPS) has been recognised as problematic for the safe operation of power systems since the 1920s [6], after the commercial use of electricity in the late 1870s with the installation of lights for lighthouses and street lighting [7]. Stability is defined as the ability of an electrical power system to return to normal or stable conditions after being disrupted [8–10]. System stability depends primarily on the behaviour of synchronous machines after a disturbance [8, 11], since the limits of system stability are characterised by the presence of a maximum amount of power passing through a point in the system without resulting in a loss of stability [11, 12]. The loss of stability in the electrical power system is strongly influenced by the presence of a number of disturbances [13], so the continuity of the electrical power system must be maintained in synchronisation conditions as perfectly as possible in a steady state [6, 12]. Disturbances in the system can be of various types [13–15], such as sudden load changes, sudden short circuits between channels and the ground [16], inter-channel disturbances, line-to-ground symmetrical disturbances, switching phenomena, and others.

Steady-state stability is essential for planning and designing power systems, developing specialised automatic control devices, placing new elements into system operation, or modifying operating conditions [17]. Stability selection is made based on the requirements of stability limits or the quality of electrical power in a steady state or during transients. The steady-state limit is guided by the maximum power flow through a given point without resulting in a loss of stability, when the power is gradually increased or decreased [18, 19]. Steady-state limit estimation is important for electrical power system analysis, with coverage on checking the electric power system in stable conditions, determining stability limits, and qualitative estimation of transient conditions [11, 17]. The existence of stability and limit estimation at steady-state is necessary for the estimation of the choice of excitation and controller system types, control modes, excitation control system parameters, and automation, including power breaker (circuit breaker, CB) performance [20, 21]. It has to do with the presence of an electrical power system made up of numerous synchronous machines that operate in unison [11, 22]. In conditions in which the system experiences interference, a force to return to a normal or stable state develops [8, 12, 23].

By simulating steady-state stability, the problem is put into words based on the system’s ability to get all networks back in sync with the same frequency after a gradual change in power. The simulation is carried out through the selection and determination of an analogy of the structure of an electric power network [24]. An analogy of the electric power structure consists of four interconnected buses with assumed values against a number of accompanying parameters. An analogy of the electrical power system structure for steady-state stability simulation [25] is shown in **Figure 1**.

Based on **Figure 1**, it is shown and can be explained that the load and connecting transmission line have impedance values, namely R + jX values (ohms). A circle with the notation "number" is a synchronous generator node in a rotating generation system. System structure with four buses, namely bus #1, bus #2, bus #3, and bus #4. Bus #1 is noted as a swing bus (slack bus, reference bus, $V - \varphi$ bus), which serves as a bearer of the lack of generation power. Buses #2 and #3 are noted as generation buses (voltage control buses, $P - V$ buses), while bus #4 is noted as a load bus ($P - Q$ bus). The next phasing is in the form of making algorithms and compiling MATLAB-based syntax, a number of simulations of the steady-state stability phenomenon with initial assumption values, and making conditions for the steady-state stability phenomenon through changes in power on the load on each bus [26, 27].

The objectives for this diagnostic are set on a pair of primary levels, referring to the formulation of the problem. First, the acquisition of the three syntax structures of the programme, namely the programme structure for inputting data on initial assumption values, the programme structure for simulating the phenomenon of power flow, and the programme structure for simulating the phenomenon of steady-state stability with data on assumption values; The second is the implementation of steady-state stability simulation with (i) data on initial assumption values and (ii) assumption data values when there are changes in load on each bus under conditions, namely a reduction of

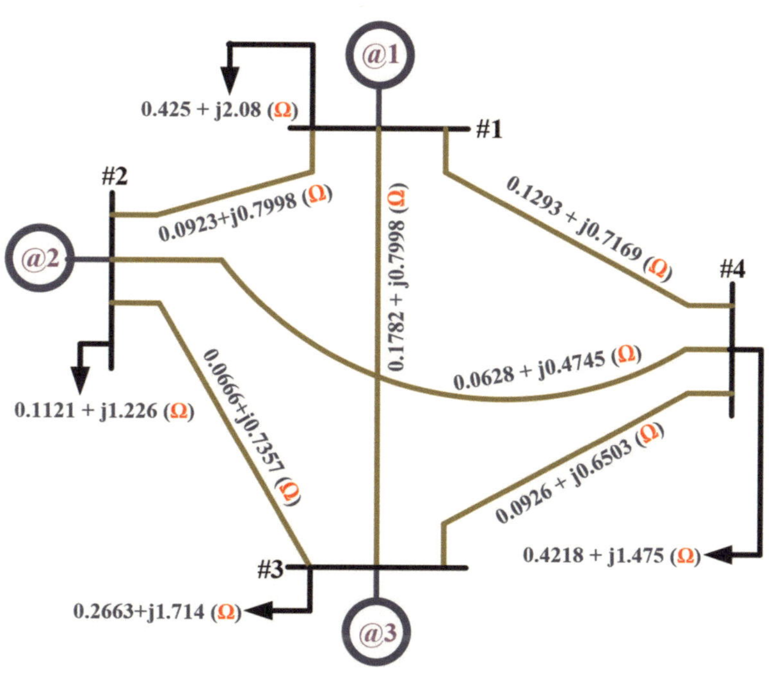

Figure 1.
Analogy of electrical power system structure for steady-state stability simulation.

power on bus #1, an addition of power on bus #2, an addition of power on bus #3, and a reduction of power on bus #4.

2. Materials and methods of simulation

2.1 Materials of simulation

The bus types for simulation consist of (a) swing buses (slack buses, reference buses, $V - \varphi$ buses), (b) generation buses (voltage control buses, $P - V$ buses), and (c) load buses ($P - Q$ buses). The swing bus is notated against bus #1, which serves as a bearer of the lack of generation power after the power flow solution is obtained. The generation bus is denoted against bus #2 and bus #3, which function as buses with active power injection parameters and the amount of bus voltage known. Load the bus for bus #4. Serves as a bus with known active power injection and reactive power bus parameters [8]. The parameters in the mathematical model are given assumed values for simulating steady-state stability parameters in the analogy of a four-bus, three-centre power system with variable changes in the value of the load (demand) power and its effect on each engine [28–30].

Guided by **Figure 1**, the initial assumption values in the model of a four-bus electric power system with three generation centres, in the form of transmission line values with a base of 100 MVA, so that units per unit are obtained. The values of the connecting transmission line components as shown in **Table 1**.

The value of the load power (demand) on each bus is shown in **Table 2**.

The value of the initial state of each generation system generator before the power flow calculation process is shown in **Table 3**.

Based on **Table 3**, it is shown and has been determined that the first bus is a swing bus, and then the generation value is adjusted after the system is operational.

Theoretically and mathematically, expressions are described in the following description: A set of rotating machines is modelled with a generator, both mechanically and electrically. Modelling the mechanical and electrical relationships of a generator, as shown in **Figure 2**.

Based on **Figure 2**, it is shown and can be explained that model block diagram electric generator on mechanical model expression is shown in Eq. (1) [8, 10, 12]:

$$T_m = J \bullet \alpha_m + T_D + T_e(\delta), \tag{1}$$

Buses		R (p.u.)	X (p.u.)
From	To		
1	2	9.23	79.98
1	3	12.93	71.69
1	4	17.82	79.98
2	3	6.28	47.45
2	4	6.66	73.57
3	4	9.26	65.03

Table 1.
The values of the connecting transmission line components.

Bus	P_d (MW)	Q_d (MVAr)	$V_{max.}$ (p.u.)	$V_{min.}$ (p.u.)
1	42.50	203.80	1.1	0.9
2	11.21	122.60	1.1	0.9
3	26.63	171.40	1.1	0.9
4	42.18	147.50	1.1	0.9

Table 2.
The value of the load power (demand) on each bus.

Bus	P_m (MW)	Q_m (MVAr)	V (p.u.)	δ (°)	H (p.u. inertia constant; sec.)	f_0 (Hz)
1	0	0	1.030	0	2.32	60
2	45	0	1.025	0	3.40	60
3	50	0	1.030	0	4.63	60

Table 3.
The value of the initial state of each generation system generator before the power flow calculation process.

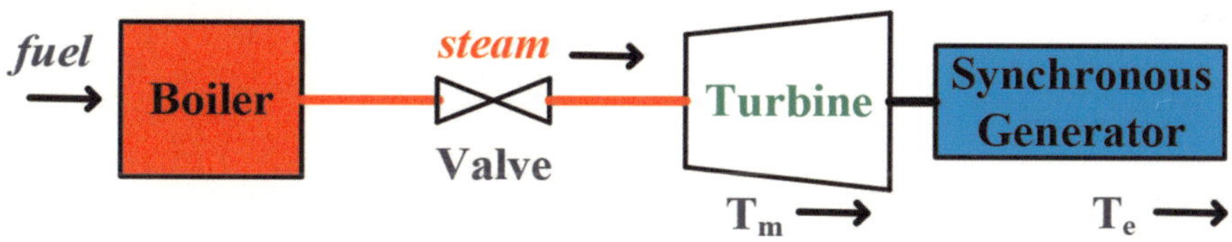

Figure 2.
Modelling the mechanical and electrical relationships of a generator.

where:
T_m = a mechanical input torque [N m];
J = a moment of inertia of the turbine and rotor [s];
α_m = an angular acceleration of the turbine and rotor [rad/s^2];
T_D = a damping torque [N m];
$T_e(\delta)$ = an equivalent electrical torque [N m].

In general, power is equal to torque multiplied by angular velocity or $P_m = T_m \bullet \omega_{sm}$. Hence, when a generator is spinning at a velocity of ω_s, Eq. (1) becomes Eq. (2).

$$P_m = T_m \bullet \omega_{sm} = [J \bullet \alpha_m + T_D + T_e(\delta)] \bullet \omega_{sm}, \tag{2}$$

which is simplified into an Eq. (3) or Eq. (4).

$$P_m = J \bullet \alpha_m \bullet \omega_{sm} + T_D \bullet \omega_{sm} + T_e(\delta) \bullet \omega_{sm}, \tag{3}$$

$$P_m = J\alpha_m \bullet \omega_s + T_D \bullet \omega_{sm} + P_e(\delta), \tag{4}$$

where, $T_e(\delta) \bullet \omega_{sm} = P_e(\delta)$.

Initially, no damping is assumed, i.e., T_D = 0, then Eq. (4) becomes Eq. (5) [8, 10, 12].

$$P_m - P_e(\delta) = J \bullet \alpha_m \bullet \omega_{sm}, \tag{5}$$

P_m is the mechanical power input, which is assumed to be constant throughout the study time period.

Regarding the rotor's angle, it is described by Eq. (6).

$$\theta_m = \omega_{sm} \bullet t + \delta_m. \tag{6}$$

where, θ_m = the rotor angle. The change in the angle of the rotor to the attitude, therefore Eq. (6), becomes Eq. (7).

$$\omega_m = \frac{d}{dt}\theta_m = \dot{\theta_m} = \frac{d}{dt}\left(\omega_{sm} \bullet t + \delta_m\right)$$

or

$$\dot{\theta_m} = \omega_{sm} + \dot{\delta_m}. \tag{7}$$

While the angular acceleration of the turbine and rotor changed with time, as explained by Eq. (8).

$$\alpha_m = \frac{d}{dt}\omega_m = \dot{\omega_m} = \ddot{\delta_m}. \tag{8}$$

The substitution of Eq. (8) to Eq. (5) is obtained like Eq. (9).

$$P_m - P_e(\delta) = J \bullet \omega_s \bullet \ddot{\delta_m}. \tag{9}$$

where $J \bullet \omega_s$ is an inertia of the machine at synchronous velocity.

Convert to per unit by dividing by the MVA rating [8, 12, 10], S_B, explained by Eq. (10).

$$\frac{P_m}{S_B} - \frac{P_e(\delta)}{S_B} = \frac{J \bullet \omega_{sm} \bullet \ddot{\delta_m}}{S_B} \bullet \frac{2\omega_{sm}}{2\omega_{sm}}. \tag{10}$$

Simplification of Eq. (10) results in Eq. (11) since $\omega_{sm} = 2 \bullet \pi \bullet f_{sm}$.

$$\frac{P_m - P_e(\delta)}{S_B} = \frac{J \bullet {\omega_{sm}}^2}{2S_B} \bullet \frac{1}{\pi \bullet f_{sm}} \bullet \ddot{\delta_m}. \tag{11}$$

Define

$$\frac{J \bullet {\omega_{sm}}^2}{2S_B} = H. \tag{12}$$

where H is a per unit inertia constant [s].

The substitution of Eq. (12) to Eq. (11) is obtained like Eq. (13).

$$\frac{P_m - P_e(\delta)}{S_B} = H \bullet \frac{1}{\pi \bullet f_{sm}} \bullet \delta_m. \tag{13}$$

All values are converted to per unit, Eq. (13) results in Eq. (14) since $M = {}^{H}\!/_{\pi \bullet f_{sm}}$ [8, 10, 12].

$$P_m - P_e(\delta) = M \bullet \ddot{\delta}_m. \tag{14}$$

Eq. (14) is known as the generator swing eq. [8, 10, 12]. Adding a damping results in Eq. (15).

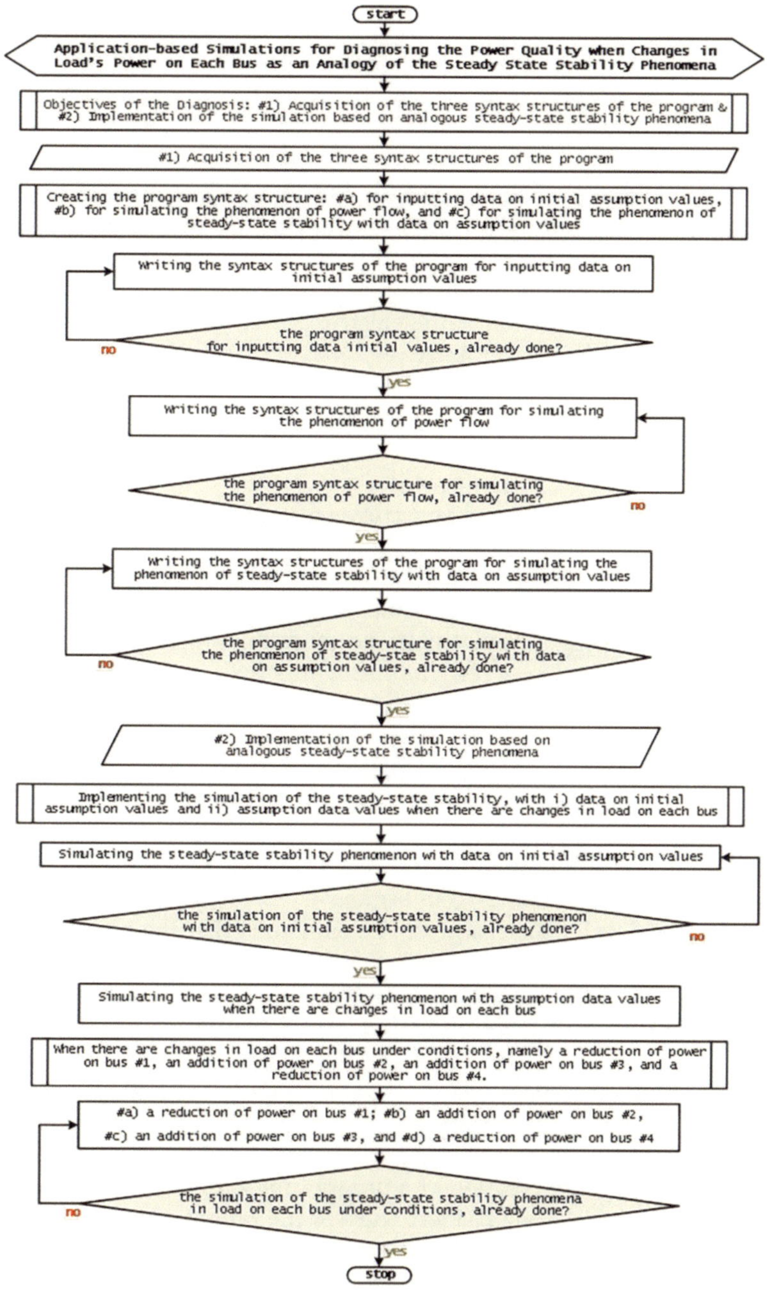

Figure 3.
The flow diagram of the simulation implementation of the existence of the steady-state stability phenomenon in the electric power system simulated with the MATLAB application.

$$P_m - P_e(\delta) = M \bullet \ddot{\delta_m} + D \bullet \dot{\delta_m}. \quad (15)$$

The simplest generator model, known as the classical model, treats the generator as a voltage source behind the direct-axis transient reactance ($\boldsymbol{X_d}'$); the voltage magnitude is fixed, but its angle changes according to the mechanical dynamics. The explanation in the form of an equation is obtained like Eq. (16) [8, 10, 12].

$$P_e(\delta) = \frac{|V_t| \bullet |E_a|}{X_d'} \sin \delta. \quad (16)$$

2.2 Methods of simulation

In order to do simulations, we can use the MATLAB application to create program structures, set up assumptions, algorithms, and syntax for the simulation process, and then run the programs to simulate the steady-state stability in the electrical power system. The flow diagram of the simulation implementation of the existence of the steady-state stability phenomenon in the electric power system simulated with the MATLAB application is shown in **Figure 3**.

3. Results and discussion

3.1 Acquisition of the three syntax structures of the programme

Acquisition of three programme syntax structures for this simulation activity, including the structure of programme syntax for data input purposes on initial assumption values, the programme syntax structure for simulating the phenomena of power flows, and the structure of programme syntax for simulating the phenomena of steady-state stabilities with data on assumption values.

3.1.1 Structuring the programme syntax for inputting data on initial assumption values

For data input of assumed values, the existence of a MATLAB-based application syntax structure of the programme is required. The programme syntax structure for the input of assumed values is shown in **Figure 4**.

Based on **Figure 4**, it is shown and can be explained that the programme syntax structure consists of (i) bus data, which includes bus number, bus type, and load power data (demand), (ii) generator data, which includes generation power, voltage, and base data, and (iii) inter-bus connexion impedance data.

3.1.2 Structuring the programme syntax for simulating the power flows phenomena

After the programme syntax structure for data entry of initial assumed values is created, it proceeds with the creation of a programme syntax structure for simulating the phenomenon of power flow. The structure of the programme syntax for simulating power flow phenomena is shown in **Figure 5**.

Based on **Figure 5**, it is shown that the programme syntax structure for simulating power flow phenomena is used to determine the mechanical power of each plant due to load changes, with operation through data input in the programme based on initial assumption values.

```
%Listing the Program Syntax Structure based on MATLAB
%FOR MULTI-MACHINE SYSTEMS
%data on initial assumption values
clc
function [baseMVA, bus, gen, branch] = case3gs
% build function for data input of bus system
%%-----  Data for Power Flow Iteration -----%%
%% system MVA base
baseMVA = 100;
%% bus data
% bus_i  type  Pd Qd Gs Bs area  Vm Va baseKV  zone  Vmax  Vmin
bus = [
      1 3  42.50     203.8   0  0  1  1  0  230   1  1.1   0.9;
      2 2  11.21     122.6      0  0  1  1  0  230   1  1.1   0.9;
      3 2  26.63     171.4      0  0  1  1  0  230   1  1.1   0.9;
      4 1  42.18     147.5      0  0  1  1  0  230   1  1.1   0.9;
];

  %% generator data
  %      bus Pg     Qg Qmax  Qmin  Vg     mBase status   Pmax  Pmin
  gen = [
         2   45     0  100   -100  1.025 100   1         0      0;
         3   50     0  100   -100  1.03  100   1         0      0;
         1   0      0  100   -100  1.03  100   1         0      0;
  ];
  %% branch data
  % fbus  tbus  r  x  b  rateA rateB rateC ratio angle status
  branch = [
     1  2  0.0923  0.7998  0.1025  250  250  250  0  0  1;
     1  4  0.1293  0.7169  0.0775  250  250  250  0  0  1;
     1  3  0.1782  0.7998  0.0775  250  250  250  0  0  1;
     2  4  0.0628  0.4745  0.0775  250  250  250  0  0  1;
     2  3  0.0666  0.7357  0.0775  250  250  250  0  0  1;
     3  4  0.0926  0.6503  0.0775  250  250  250  0  0  1;
  ];
  return;
```

Figure 4.
The programme syntax structure for the input of assumed values.

After obtaining data on the mechanical power value of each power generation machine, it is necessary to calculate the power flow with data on the values of the initial assumption. The initial mechanical power value is calculated with the programme syntax structure for the power flow phenomenon shown in **Table 4**.

3.1.3 Programme syntax structure for simulating the phenomena of steady-state stabilities

The final stage of creating a programme syntax structure that will be used to simulate steady-state stability through changes in load values on each bus. The structure of the programme syntax for simulating the phenomena of steady-state stability is shown in **Figure 6**.

Based on **Figure 6**, it can be seen that the programme syntax structure for the steady-state stability phenomenon is made up of (i) mechanical power values as input data from power flow programmes; (ii) data on each machine in the form of voltage value, moment of inertia, dumping ratio, frequency, and resistance in the machine; and (iii) equations for finding the current, voltage, frequency, rotor angle, rotor angle delta, and angular velocity.

3.2 Simulation results

3.2.1 The result of the simulation on the phenomenon of steady-state stability uses the data from the initial assumption values

Some curves of the result of the simulation on the phenomenon of steady-state stability using the data from the initial assumption values are shown in **Figure 7**.

```
%Listing the Program Syntax Structure based on MATLAB
%FOR MULTI-MACHINE SYSTEMS
%simulating the phenomena of power flows
clc
function [MVAbase, bus, gen, branch, success, et] = ...
    runpf(casename, mpopt, fname, solvedcase)
 [PQ, PV, REF, NONE, BUS_I, BUS_TYPE, PD, QD, GS, BS, BUS_AREA, VM, ...
    VA, BASE_KV, ZONE, VMAX, VMIN, LAM_P, LAM_Q, MU_VMAX, MU_VMIN] = idx_bus;
[F_BUS, T_BUS, BR_R, BR_X, BR_B, RATE_A, RATE_B, ...
    RATE_C, TAP, SHIFT, BR_STATUS, PF, QF, PT, QT, MU_SF, MU_ST] = idx_brch;
[GEN_BUS, PG, QG, QMAX, QMIN, VG, MBASE, ...
    GEN_STATUS, PMAX, PMIN, MU_PMAX, MU_PMIN, MU_QMAX, MU_QMIN] = idx_gen;
%% default arguments
if nargin < 4
    solvedcase = '';   %% don't save solved case
    if nargin < 3
        fname = '';    %% don't print results to a file
        if nargin < 2
            mpopt = mpoption;   %% use default options
            if nargin < 1
                casename = 'case9'; %% default data file is 'case9.m'
            end
        end
    end
end
%% options
verbose = mpopt(31);
qlim = mpopt(6);            %% enforce Q limits on gens?
dc = mpopt(10);             %% use DC formulation?
%% read data & convert to internal bus numbering
[baseMVA, bus, gen, branch] = loadcase(casename);
[i2e, bus, gen, branch] = ext2int(bus, gen, branch);
%% get bus index lists of each type of bus
[ref, pv, pq] = bustypes(bus, gen);
%% generator info
on = find(gen(:, GEN_STATUS) > 0);   %% which generators are on?
gbus = gen(on, GEN_BUS);             %% what buses are they at?
%%-----  run the power flow  -----
t0 = clock;

if dc  %% DC formulation
    %% initial state
    Va0 = bus(:, VA) * (pi/180);
        %% build B matrices and phase shift injections
    [B, Bf, Pbusinj, Pfinj] = makeBdc(baseMVA, bus, branch);
        %% compute complex bus power injections (generation - load)
    %% adjusted for phase shifters and real shunts
    Pbus = real(makeSbus(baseMVA, bus, gen)) - Pbusinj - bus(:, GS) / baseMVA;
        %% "run" the power flow
    Va = dcpf(B, Pbus, Va0, ref, pv, pq);
        %% update data matrices with solution
    branch(:, [QF, QT]) = zeros(size(branch, 1), 2);
    branch(:, PF) = (Bf * Va + Pfinj) * baseMVA;
    branch(:, PT) = -branch(:, PF);
    bus(:, VM) = ones(size(bus, 1), 1);
    bus(:, VA) = Va * (180/pi);
    %% update Pg for swing generator (note: other gens at ref bus are accounte
for in Pbus)
    %%      Pg = Pinj + Pload + Gs
    %%      newPg = oldPg + newPinj - oldPinj
    refgen = find(gbus == ref);              %% which is(are) the reference gen
    gen(on(refgen(1)), PG) = gen(on(refgen(1)), PG) + (B(ref, :) * Va - Pbus(r
* baseMVA;
        success = 1;
else                    %% AC formulation
    %% initial state
    % V0    = ones(size(bus, 1), 1);      %% flat start
    V0  = bus(:, VM) .* exp(sqrt(-1) * pi/180 * bus(:, VA));
    V0(gbus) = gen(on, VG) ./ abs(V0(gbus)).* V0(gbus);
        if qlim
        ref0 = ref;                 %% save index and angle of
```

```
        Varef0 = bus(ref0, VA); %%   original reference bus
        limited = [];           %% list of indices of gens @ Q lims
        fixedQg = zeros(size(gen, 1), 1);   %% Qg of gens at Q limits
    end

 repeat = 1;
    while (repeat)
      %% build admittance matrices
      [Ybus, Yf, Yt] = makeYbus(baseMVA, bus, branch);
      %% compute complex bus power injections (generation - load)
      Sbus = makeSbus(baseMVA, bus, gen);
      %% run the power flow
      alg = mpopt(1);
      if alg == 1
      [V, success, iterations] = newtonpf(Ybus, Sbus, V0, ref, pv, pq, mpopt);
      elseif alg == 2 | alg == 3
      [Bp, Bpp] = makeB(baseMVA, bus, branch, alg);
      [V, success, iterations] = fdpf(Ybus, Sbus, V0, Bp, Bpp, ref, pv, pq, mpopt)
      elseif alg == 4
      [V, success, iterations] = gausspf(Ybus, Sbus, V0, ref, pv, pq, mpopt);
      else
      error('Only Newton''s method, fast-decoupled, and Gauss-Seidel power flow
 algorithms currently implemented.');
        end
      %% update data matrices with solution
      [bus, gen, branch] = pfsoln(baseMVA, bus, gen, branch, Ybus, Yf, Yt, V, ref,
 pv, pq);
       if qlim         %% enforce generator Q limits
       %% find gens with violated Q constraints
         mx = find( gen(:, GEN_STATUS) > 0 & gen(:, QG) > gen(:, QMAX) );
         mn = find( gen(:, GEN_STATUS) > 0 & gen(:, QG) < gen(:, QMIN) );
       if ~isempty(mx) | ~isempty(mn)  %% we have some Q limit violations
         if verbose & ~isempty(mx)
          fprintf('Gen %d at upper Q limit, converting to PQ bus\n', mx);
         end
         if verbose & ~isempty(mn)
         fprintf('Gen %d at lower Q limit, converting to PQ bus\n', mn);
         end
  %% save corresponding limit values
        fixedQg(mx) = gen(mx, QMAX);
        fixedQg(mn) = gen(mn, QMIN);
        mx = [mx;mn];
        %% convert to PQ bus
        gen(mx, QG) = fixedQg(mx);   %% set Qg to binding limit
        gen(mx, GEN_STATUS) = 0;     %% temporarily turn off gen,
        for i = 1:length(mx)         %% (one at a time, since
        bi = gen(mx(i), GEN_BUS);   %%  they may be at same bus)
        bus(bi, [PD,QD]) = ...      %% adjust load accordingly,
        bus(bi, [PD,QD]) - gen(mx(i), [PG,QG]);
        end
        bus(gen(mx, GEN_BUS), BUS_TYPE) = PQ; %% & set bus type to PQ
    %% update bus index lists of each type of bus
       ref_temp = ref;
       [ref, pv, pq] = bustypes(bus, gen);
       if verbose & ref ~= ref_temp
       fprintf('Bus %d is new slack bus\n', ref);
       end
       limited = [limited; mx];
         else
           repeat = 0; %% no more generator Q limits violated
           end
         else
         repeat = 0;%% don't enforce generator Q limits, once is enough
         end
    end
 if qlim & ~isempty(limited)
         %% restore injections from limited gens (those at Q limits)
         gen(limited, QG) = fixedQg(limited);    %% restore Qg value,
         for i = 1:length(limited)          %% (one at a time, since
    bi = gen(limited(i), GEN_BUS);  %%  they may be at same bus)
    bus(bi, [PD,QD]) = ...           %% re-adjust load,
                bus(bi, [PD,QD]) + gen(limited(i), [PG,QG]);
    end

  gen(limited, GEN_STATUS) = 1;      %% and turn gen back on
```

```
        if ref ~= ref0
          %% adjust voltage angles to make original ref bus correct
      bus(:, VA) = bus(:, VA) - bus(ref0, VA) + Varef0;
            end
         end
    end
    et = etime(clock, t0);
    %%-----  output results  -----
    %% convert back to original bus numbering & print results
    [bus, gen, branch] = int2ext(i2e, bus, gen, branch);
    if fname
      [fd, msg] = fopen(fname, 'at');
      if fd == -1
        error(msg);
      else
       printpf(baseMVA, bus, gen, branch, [], success, et, fd, mpopt);
        fclose(fd);
         end
    end
    printpf(baseMVA, bus, gen, branch, [], success, et, 1, mpopt);

    %% save solved case
    if solvedcase
        savecase(solvedcase, baseMVA, bus, gen, branch);
    end
    %% this is just to prevent it from printing baseMVA
    %% when called with no output arguments
    if nargout, MVAbase = baseMVA; end
    return;
```

Figure 5.
The structure of the programme syntax for simulating power flow phenomena.

Bus	Voltages		Power from synchronous generator		Power to Loads	
	Mag. (p.u.)	Angle (°)	P_e (MW)	Q_e (MVAr)	P_d (MW)	Q_d (MVAr)
1	1.030	0.00	35.43	258.06	42.50	203.8
2	1.025	5.464	45.00	176.20	11.21	122.6
3	1.030	3.912	50.00	239.05	26.63	171.4
4	0.514	−0.723	—	—	42.18	147.5
		Total:	130.43	673.31	122.52	645,.0

Table 4.
The initial mechanical power value is calculated with the programme syntax structure for the power flow phenomenon.

Based on **Figure 7**, it is shown and can be explained that the time for achieving stable conditions is preceded by machine #1, then machine #2, and finally machine #3 for all parameters. The values of a number of parameters are simulated with data assumption values, as shown in **Table 5**.

Based on **Table 5**, it is shown and can be explained that there are a number of phenomena, namely (a) changes in rotor angle values when achieving stable, (b) achieving time when delta rotor reaches 0°, achieving stable, (c) achieving time when delta omega values reach 0 rad/s, achieving stable, and (d) achieving time when frequency values reach 60 hertz, achieving stable. The generating engine with the largest power generation (in this case, machine #3) takes longer to achieve a state of stability again after a change in load.

```
%Listing the Program Syntax Structure based on MATLAB
%FOR MULTI-MACHINE SYSTEMS
%simulating the phenomena of steady-stte stabilities
clc
% Data input
% Vbus H X D Fo
V1=1.030; H1=2.32; X1=0.32; D1=0.4; f01=60; %machine #1
V2=1.025; H2=3.40; X2=0.33; D2=0.4; f02=60; %machine #2
V3=1.030; H3=4.63; X3=0.20; D3=0.4; f03=60; %machine #3
%mechanical power calculation results after power flow estimation
S1=input ('Enter the value of the mechanical power of Machine #1 = ')/100;
S2=input ('Enter the value of the mechanical power of Machine #2 = ')/100;
S3=input ('Enter the value of the mechanical power of Machine #3 = ')/100;

%the real value of mechanical power
Pm1=abs(S1);
Pm2=abs(S2);
Pm3=abs(S3);
%Calculation of the electric current produced by each machine
I1=conj(S1)/conj(V1) %on the machine #1
I2=conj(S2)/conj(V2) %on the machine #2
I3=conj(S2)/conj(V2) %on the machine #3
%calculation of the output voltage of each machine
E1=V1+I1*X1 %calculation of the output voltage on the machine #1
E2=V2+I2*X2 %calculation of the output voltage on the machine #2
E3=V3+I3*X3 %calculation of the output voltage on the machine #3
%calculation of rotation changes and frequency on machine #1
disp(' Execution Results')
Pmax1=abs(E1)*V1/X1, d01=asin(Pm1/Pmax1)        %the maximum power
Ps1=Pmax1*cos(d01)                          %synchronization power coefficient
wn1=sqrt(pi*60/H1*Ps1)                      %undamped frequency of oscillation
z1=D1/2*sqrt(pi*60/(H1*Ps1))                 %damping ratio
wd1=wn1*sqrt(1-z1^2),fd1=wd1/(2*pi)           %damped frequency oscilation
tau1=1/(z1*wn1)                             %fixed time
th1=acos(z1)                               %fixed phase angle
Dd01=10*pi/180                            %initial angle in radian
t1=0:.01:1;                                 %time interval
Dd1=Dd01/sqrt(1-z1^2)*exp(-z1*wn1*t1).*sin(wd1*t1+th1); %change in rotor angle
d1=(d01+Dd1)*180/pi ;                      %power angle in degree
Dw1=-wn1*Dd01/sqrt(1-z1^2)*exp(-z1*wn1*t1).*sin(wd1*t1); %change in rotor angle and
rotation speed
f1=f01+Dw1/(2*pi);             %frequency (Hz)
%calculation of rotation changes and frequency on machine #2
Pmax2=abs(E2)*V2/X2, d02=asin(Pm2/Pmax2)        %the maximum power
Ps2=Pmax2*cos(d02)                         % synchronization power coefficient
wn2=sqrt(pi*60/H2*Ps2)                      %undamped frequency of oscillation
z2=D2/2*sqrt(pi*60/(H2*Ps2))                 %damping ratio
wd2=wn2*sqrt(1-z2^2),fd2=wd2/(2*pi)           %damped frequency oscill
tau2=1/(z2*wn2)                             %fixed time
th2=acos(z2)                               %fixed phase angle
Dd02=10*pi/180                            %initial angle in radian
Dd2=Dd02/sqrt(1-z2^2)*exp(-z2*wn2*t1).*sin(wd2*t1+th2);%change in rotor angle
d2=(d02+Dd2)*180/pi ;                       %power angle in degree
Dw2=-wn2*Dd02/sqrt(1-z2^2)*exp(-z2*wn2*t1).*sin(wd2*t1);%change in rotor angle and
rotation speed
f2=f02+Dw2/(2*pi) ;              %frequency (Hz)
%calculation of rotation changes and frequency on machine #3
Pmax3=abs(E3)*V3/X3, d03=asin(Pm3/Pmax3)   %the maximum power
Ps3=Pmax3*cos(d03)                      %Koefisien Daya Sinkron
wn3=sqrt(pi*60/H3*Ps3)             %undamped frequency of oscillation
z3=D3/2*sqrt(pi*60/(H3*Ps3))                 %damping ratio
wd3=wn3*sqrt(1-z3^2),fd3=wd3/(2*pi)           %damped frequency oscill
tau3=1/(z3*wn3)                            %fixed time
th3=acos(z3)                               %fixed phase angle
Dd03=10*pi/180                            %initial angle in radian
Dd3=Dd03/sqrt(1-z3^2)*exp(-z3*wn3*t1).*sin(wd3*t1+th3); %change in rotor angle
d3=(d03+Dd3)*180/pi   ;                     %power angle in degree
xlabel('time, t (sec)'), ylabel('Delta Omega in (rad/sec)')
subplot(4,1,1), plot(t1,d1,t1,d2,t1,d3), grid %curving the rotor angle change on machine #3
xlabel('time, t (sec)'), ylabel('Rotor Angle in (degree)')
subplot(111)
disp('**************************curve #1****************************')
disp('')
disp('******************** END **********************')
```

Figure 6.
The structure of the programme syntax for simulating the phenomena of steady-state stability.

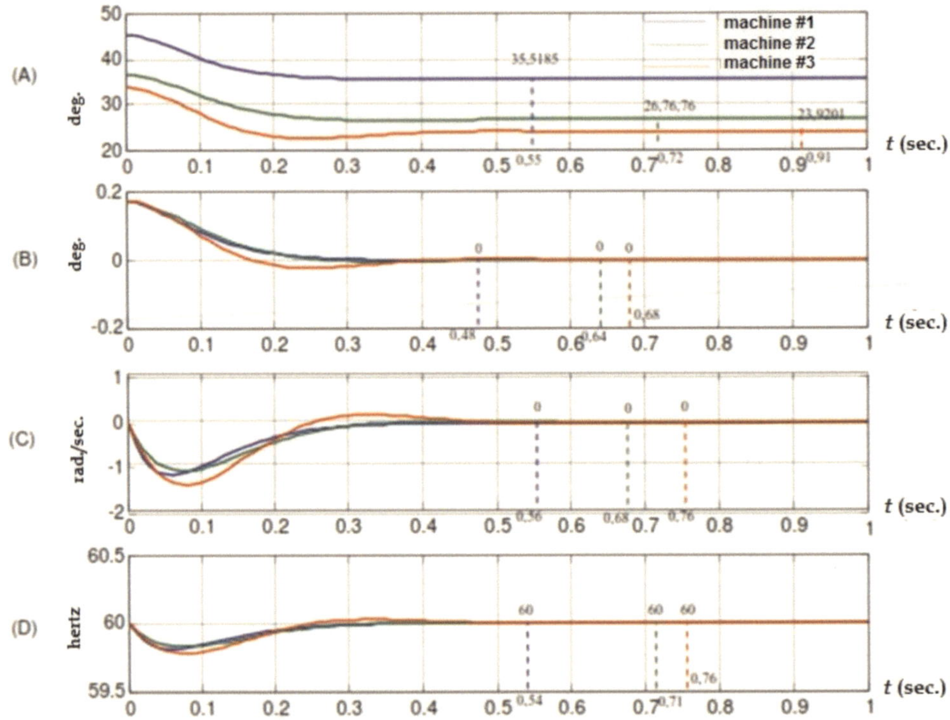

Figure 7.
Some curves of the result of the simulation on the phenomenon of steady-state stability using the data from the initial assumption values. (A) rotor angle curves (B) rotor angle difference curves (C) rotor angular velocity difference curves (D) frequency curves.

3.2.2 The result of the simulation on the phenomenon of steady-state stability uses the data from the assumption values for load changes on each bus

Depending on the needs of the user, today's linked power networks may supply a range of loads. Naturally, these demands change continually, which causes the system's loading to vary and has a variety of effects [26, 27]. Variation in loading has certain negative impacts, with the following listed as the most notable, namely (i) costlier power generation, (ii) complex system control, (iii) a need for more equipment, and (iv) higher losses are just a few of the consequences [27–29]. The reduction or addition of power due to the release or inclusion of a number of type and range of loads [25, 28–30]. In the simulation of the steady-state stability phenomenon, it is made with four conditions: (a) power reduction on bus #1, (b) power addition on bus #2, (c) power addition on bus #3, and (d) power reduction on bus #4.

3.2.2.1 The simulation on the phenomenon of steady-state stability uses the data of power reduction on bus #1

Changes in load power values, in the form of reducing active and reactive power from P_d = 42.50 MW and Q_d = 203.8 MVAr to P_d = 25.00 MW and Q_d = 125.5 MVAr. The value of the load power of each bus after the reduction of power on bus #1 is shown in **Table 6**.

Parameters	Machines		
	#1	#2	#3
(A) Rotor angle			
The initial value unstable (°)	45.5198	36.7641	33.9222
Stable value (°)	35.5185	26.7676	23.9201
Stable time achievement (s)	0.55	0.72	0.91
(B) Rotor angle difference			
The initial value unstable (°)	0.1745	0.1745	0.1745
Stable value (°)	0.0000	0.0000	0.0000
Stable time achievement (s)	0.48	0.64	0.68
(C) Rotor angular velocity difference			
The initial value unstable (rad/s)	−0.4400	−0.3119	−0.359
Stable value (rad/s)	0.0000	0.0000	0.0000
Stable time achievement (s)	0.56	0.68	0.76
(D) Frequency			
The initial value unstable (Hz)	59.9300	59.9504	59.9428
Stable value (Hz)	60.0000	60.0000	60.0000
Stable time achievement (s)	0.54	0.71	0.76

Table 5.
The values of a number of parameters are simulated with data assumption values.

Bus	P_d (MW)	Q_d (MVAr)	$V_{max.}$ (p.u.)	$V_{min.}$ (p.u.)
1	***25.00***	***125.5***	1.1	0.9
2	11.21	122.6	1.1	0.9
3	26.63	171.4	1.1	0.9
4	42.18	147.5	1.1	0.9

Table 6.
The value of the load power of each bus after the reduction of power on bus #1.

The values in **Table 6** are used for mechanical power gain after a change in power values on bus #1. The mechanical power values simulated with the power flow programme with the case of power change on bus #1 are shown in **Table** 7.

Some curves from the simulation of steady-state stability with power changes on bus #1 are shown in **Figure 8**.

Based on **Figure 8**, it is shown and can be explained that the time for achieving stable conditions is preceded by machine #1, then machine #2, and finally machine #3 for all parameters.

When the active power and reactive power at the connected load on bus #1 went from P_d = 42.50 MW and Q_d = 203.8 MVAr to P_d = 25.00 MW and Q_d = 125.5 MVAr, a number of things started to happen. A number of phenomena due to power reduction on bus #1 are shown in **Table 8**.

Bus	Voltages		Power from the synchronous generator		Power to loads	
	Mag. (p.u.)	Ang. (°)	P_e (MW)	Q_e (MVAr)	P_d (MW)	Q_d (MVAr)
1	1.030	0.000	17.93	179.76	25.50	125.50
2	1.025	5.464	45.00	176.20	11.21	122.60
3	1.030	3.912	50.00	239.05	26.63	171.40
4	0.514	−0.723	—	—	42.18	147.50
	Total value:		112.93	595.01	105.02	567.00

Table 7.
The mechanical power values simulated with the power flow programme with the case of power change on bus #1.

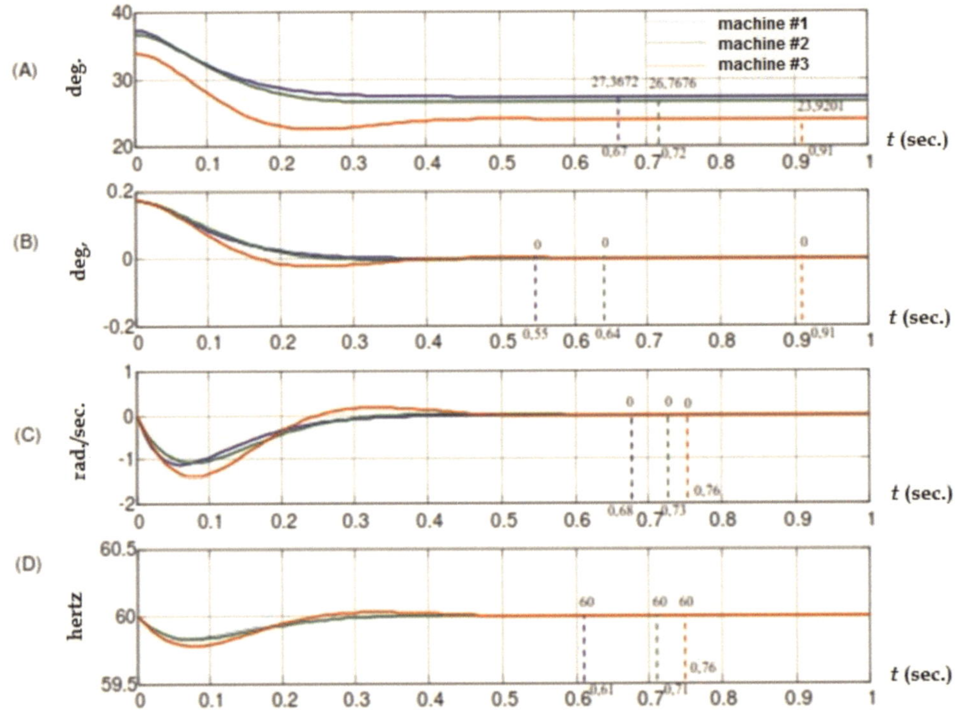

Figure 8.
Some curves from the simulation of steady-state stability with power changes on bus #1. (A) Rotor angle curves, (B) rotor angle difference curves, (C) rotor angular velocity difference curves and (D) frequency curves.

3.2.2.2 The simulation on the phenomenon of steady-state stability uses the data of power addition on bus #2

Changes in load power values, in the form of adding active and reactive power from P_d = 11.21 MW and Q_d = 126.6 MVAr to P_d = 40.00 MW and Q_d = 200 MVAr. The value of the load power of each bus after the addition of power on bus #2 is shown in **Table 9**.

The values in **Table 9** are used for mechanical power gain after a change in power values on bus #2. The mechanical power values simulated with the power flow programme with the case of power change on bus #2 are shown in **Table 10**.

Phenomena	Mach.	Initial value (°)	Final value (°)	Difference (°)	Information
Changes in rotor angle value when stable conditions are reached	#1	35.5198	27.3672	8.1526	Smaller
	#2	26.7676	26.7676	0	Fixed value
	#3	23.9201	23.9201	0	Fixed value
Achievement of the time when the rotor delta is stable (at a value of 0°)	#1	0.48	0.55	0.07	Slower
	#2	0.64	0.64	0	Fixed value
	#3	0.68	0.91	0.23	Slower
Achievement of time when delta omega value reaches 0 rad/s (stable)	#1	0.56	0.68	0.12	Slower
	#2	0.68	0.73	0.05	Slower
	#3	0.76	0.76	0	Fixed value
Achievement of time when the frequency value reached 60 Hz (stable)	#1	0.54	0.61	0.07	Slower
	#2	0.71	0.71	0	Slower
	#3	0.76	0.76	0	Fixed value

Table 8.
A number of phenomena due to power reduction on bus #1.

Bus	P_d (MW)	Q_d (MVAr)	$V_{max.}$ (p.u.)	$V_{min.}$ (p.u.)
1	25.00	125.50	1.1	0.9
2	***40.00***	***200.00***	***1.1***	***0.9***
3	26.63	171.40	1.1	0.9
4	42.18	147.50	1.1	0.9

Table 9.
The value of the load power of each bus after the addition of power on bus #2.

Bus	Voltages		Power from the synchronous generator		Power to loads	
	Mag. (p.u.)	Ang. (°)	P_e (MW)	Q_e (MVAr)	P_d (MW)	Q_d (MVAr)
1	1.030	0.00	73.60	275.64	42.50	203.80
2	1.025	−2.137	45.00	282.56	40.00	200.00
3	1.030	−1.120	50.00	269.23	26.63	171.40
4	0.325	−0.723	—	—	42.18	147.50
	Total value:		168.60	827.43	151.31	722.70

Table 10.
The mechanical power values simulated with the power flow programme with the case of power change on bus #2.

Some curves from the simulation of steady-state stability with power changes on bus #2 are shown in **Figure 9**.

Based on **Figure 9**, it is shown and can be explained that the time for achieving stable conditions is preceded by machine #1, then machine #2, and finally machine #3 for all parameters.

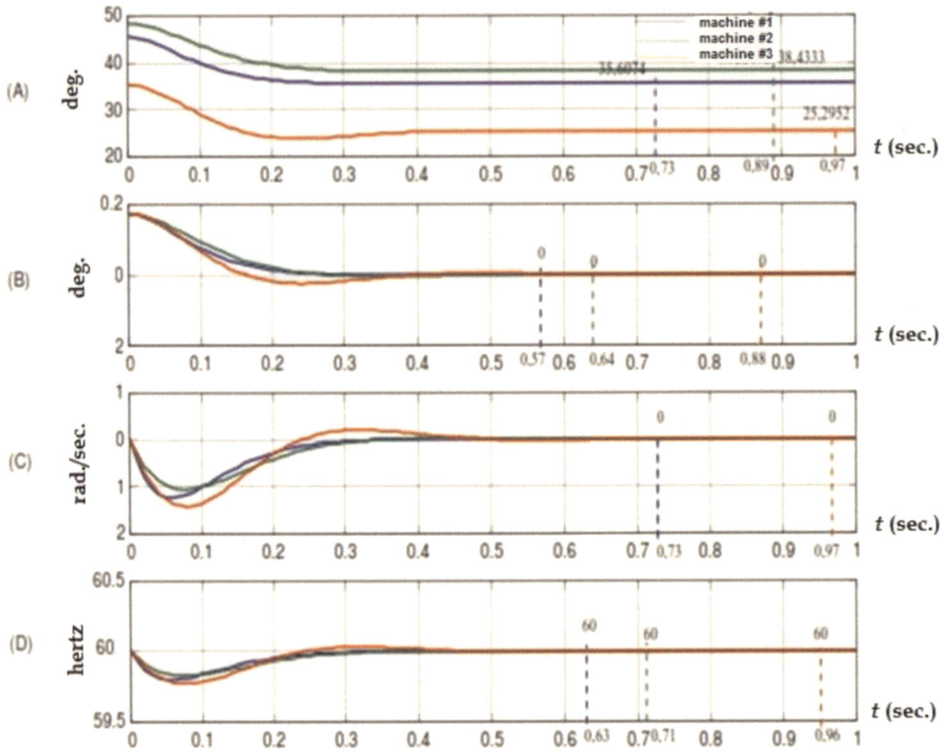

Figure 9.
Some curves from the simulation of steady-state stability with power changes on bus #2. (A) Rotor angle curves, (B) rotor angle difference curves, (C) rotor angular velocity difference curves and (D) frequency curves.

When the active power and reactive power at the connected load on bus #2 went from P_d = 11.21 MW and Q_d = 122.6 MVAr to P_d = 40.00 MW and Q_d = 200.00 MVAr, a number of things started to happen. A number of phenomena due to power reduction on bus #2 are shown in **Table 11**.

3.2.2.3 The simulation on the phenomenon of steady-state stability uses the data of power addition on bus #3

Changes in load power values, in the form of adding active and reactive power from P_d = 26.63 MW and Q_d = 171.4 MVAr to P_d = 35.75 MW and Q_d = 195.5 MVAr. The value of the load power of each bus after the addition of power on bus #2 is shown in **Table 9**.

The values in **Table 12** are used for mechanical power gain after a change in power values on bus #3. The mechanical power values simulated with the power flow programme with the case of power change on bus #3 are shown in **Table 13**.

Some curves from the simulation of steady-state stability with power changes on bus #3 are shown in **Figure 10**.

Based on **Figure 10**, it is shown and can be explained that the time for achieving stable conditions is preceded by machine #1, then machine #2, and finally machine #3 for all parameters.

Phenomena	Mach.	Initial value (°)	Final value (°)	Difference (°)	Information
Changes in rotor angle value when stable conditions are reached	#1	35.5198	35.6074	0.0876	Greater
	#2	26.7676	38.4333	11.6657	Greater
	#3	23.9201	25.2952	1.3751	Greater
Achievement of the time when the rotor delta is stable (at a value of 0°)	#1	0.48	0.57	0.09	Slower
	#2	0.64	0.64	0	Fixed value
	#3	0.68	0.88	0.20	Slower
Achievement of time when delta omega value reaches 0 rad/s (stable)	#1	0.56	0.73	0.17	Slower
	#2	0.68	0.73	0.05	Slower
	#3	0.76	0.97	0.21	Slower
Achievement of time when the frequency value reached 60 HZ (stable)	#1	0.54	0.63	0.09	Slower
	#2	0.71	0.71	0	Fixed value
	#3	0.76	0.96	0.20	Slower

Table 11.
A number of phenomena due to power reduction on bus #2.

Bus	P_d (MW)	Q_d (MVAr)	$V_{max.}$ (p.u.)	$V_{min.}$ (p.u.)
1	25.00	125.50	1.1	0.9
2	40.00	200.00	1.1	0.9
3	***35.75***	***195.50***	***1.1***	***0.9***
4	42.18	147.50	1.1	0.9

Table 12.
The value of the load power of each bus after the addition of power on bus #3.

Bus	Voltages		Power from the synchronous generator		Power to loads	
	Mag. (p.u.)	Ang. (°)	P_e (MW)	Q_e (MVAr)	P_d (MW)	Q_d (MVAr)
1	1.030	0.00	64.40	303.52	42.50	203.80
2	1.025	5.464	45.00	231.56	11.21	122.60
3	1.030	3.912	50.00	328.09	35.75	195.50
4	0.114	−0.723	—	—	42.18	147.50
	Total value:		159.40	863.17	131.64	669.40

Table 13.
The mechanical power values simulated with the power flow programme with the case of power change on bus #3.

When the active power and reactive power at the connected load on bus #3 went from P_d = 26.63 MW and Q_d = 171.4 MVAr to P_d = 35.75 MW and Q_d = 195.5 MVAr, a number of things started to happen. A number of phenomena due to power reduction on bus #3 are shown in **Table 14**.

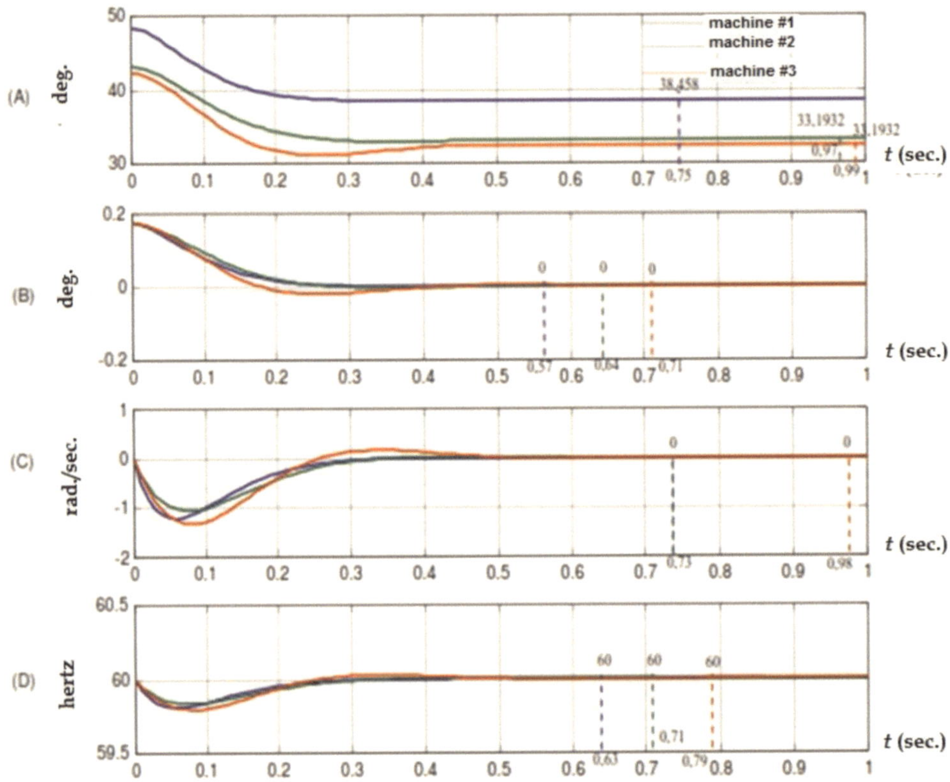

Figure 10.
Some curves from the simulation of steady-state stability with power changes on bus #3. (A) Rotor angle curves, (B) rotor angle difference curves, (C) rotor angular velocity difference curves and (D) frequency curves.

3.2.2.4 The simulation on the phenomenon of steady-state stability uses the data of power reduction on bus #4

Changes in load power values, in the form of reducing active and reactive power from P_d = 42.18 MW and Q_d = 147.5 MVAr to P_d = 10.15 MW and Q_d = 60.75 MVAr. The value of the load power of each bus after the reduction of power on bus #1 is shown in **Table 15**.

The values in **Table 15** are used for mechanical power gain after a change in power values on bus #4. The mechanical power values simulated with the power flow programme with the case of power change on bus #4 are shown in **Table 16**.

Some curves from the simulation of steady-state stability with power changes on bus #4 are shown in **Figure 11**.

Based on **Figure 11**, it is shown and can be explained that the time for achieving stable conditions is preceded by machine #1, then machine #2, and finally machine #3 for all parameters.

When the active power and reactive power at the connected load on bus #4 went from P_d = 42.18 MW and Q_d = 147.5 MVAr to P_d = 10.15 MW and Q_d = 60.75 MVAr., a number of things started to happen. A number of phenomena due to power reduction on bus #4 are shown in **Table 17**.

Phenomena	Mach.	Initial value (°)	Final value (°)	Difference (°)	Information
Changes in rotor angle value when stable conditions are reached	#1	35.5198	38.4580	2.9382	Greater
	#2	26.7676	33.1932	6.4256	Greater
	#3	23.9201	32.3113	8.912	Greater
Achievement of the time when the rotor delta is stable (at a value of 0°)	#1	0.48	0.57	0.09	Slower
	#2	0.64	0.64	0	Fixed value
	#3	0.68	0.71	0.03	Slower
Achievement of time when delta omega value reaches 0 rad/s (stable)	#1	0.56	0.73	0.17	Slower
	#2	0.68	0.73	0.05	Slower
	#3	0.76	0.98	0.22	Slower
Achievement of time when the frequency value reached 60 Hz (stable)	#1	0.54	0.63	0.09	Slower
	#2	0.71	0.71	0	Fixed value
	#3	0.76	0.79	0.03	Slower

Table 14.
A number of phenomena due to power reduction on bus #3.

Bus	P_d (MW)	Q_d (MVAr)	$V_{max.}$ (p.u.)	$V_{min.}$ (p.u.)
1	25.00	125.50	1.1	0.9
2	40.00	200.00	1.1	0.9
3	35.75	195.50	1.1	0.9
4	*10.15*	*60.75*	*1.1*	*0.9*

Table 15.
The value of the load power of each bus after the reduction of power on bus #4.

Bus	Voltages		Power from the synchronous generator		Power to loads	
	Mag. (p.u.)	Ang. (°)	P_e (MW)	Q_e (MVAr)	P_d (MW)	Q_d (MVAr)
1	1.030	0.00	−2.18	219.65	42.50	203.80
2	1.025	8.836	45.00	124.71	11.21	122.60
3	1.030	7.622	50.00	180.21	26.63	171.40
4	0.887	5.360	—	—	10.15	60.75
	Total value:		92.82	524.57	90.49	558.55

Table 16.
The mechanical power values simulated with the power flow programme with the case of power change on bus #4.

3.2.2.5 Summary of the simulation results of the steady-state stability phenomenon with data on changes in load power values on each bus

Referring to a number of tables, including **Tables 8, 11, 14** and **17** are shown and can be explained that there are a number of phenomena, namely (a) changes in rotor

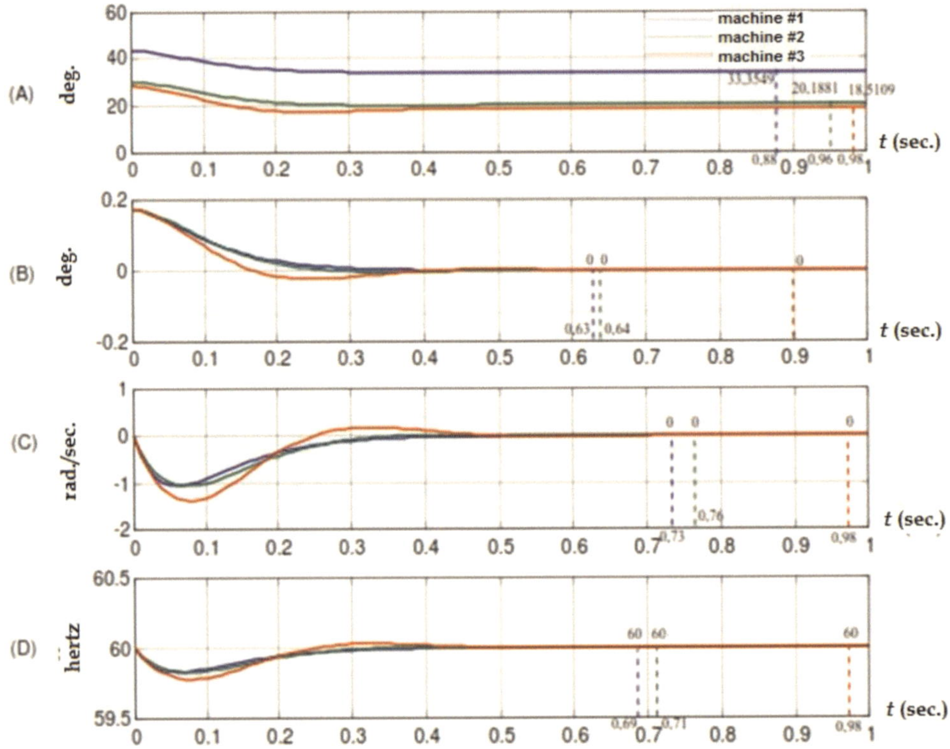

Figure 11.
Some curves from the simulation of steady-state stability with power changes on bus #4. (A) Rotor angle curves, (B) rotor angle difference curves, (C) rotor angular velocity difference curves and (D) frequency curves.

Phenomena	Mach.	Initial value (°)	Final value (°)	Difference (°)	Information
Changes in rotor angle value when stable conditions are reached	#1	35.5198	33.3549	2.1649	Smaller
	#2	26.7676	20.1882	6.5794	Smaller
	#3	23.9201	18.5109	5.4092	Smaller
Achievement of the time when the rotor delta is stable (at a value of 0°)	#1	0.48	0.63	0.15	Slower
	#2	0.64	0.64	0	Fixed value
	#3	0.68	0.90	0.22	Slower
Achievement of time when delta omega value reaches 0 rad/s (stable)	#1	0.56	0.73	0.17	Slower
	#2	0.68	0.76	0.08	Slower
	#3	0.76	0.98	0.22	Slower
Achievement of time when the frequency value reached 60 hertz (stable)	#1	0.54	0.69	0.15	Slower
	#2	0.71	0.71	0	Fixed value
	#3	0.76	0.98	0.22	Slower

Table 17.
A number of phenomena due to power reduction on bus #4.

angle values when achieving stable, (b) achieving time when the rotor delta reaches 0° (stable), (c) achieving time when the omega delta value reaches 0 rad/s (stable), and (d) achieving time when the frequency value reaches 60 hertz (stable). The generating engine with the largest power generation (in this case, machine #3) takes longer to achieve a state of stability again after a change in load.

4. Conclusions

Based on a number of descriptions in the discussion, conclusions were drawn according to the purpose of the study. The programme syntax structure consists of (i) bus data, which includes bus number, bus type, and load power data (demand); (ii) generator data, which includes generation power, voltage value, and base data; and (iii) inter-bus connection impedance data; The programme syntax structure for simulating power flow phenomena is used for the determination of the mechanical power of each power generation machine due to load changes, with operation through data input in the programme input data of initial assumption values. The initial mechanical power values inputted to the programme syntax structure for the steady-state stability phenomenon were 35.43 MW for machine #1, 45 MW for machine #2, and 50 MW for machine #3. The programme syntax structure for the steady-state stability phenomenon consists of (i) mechanical power values as input data obtained from the results of calculating the power flow programme; (ii) data on each machine in the form of voltage value, moment of inertia, dumping ratio, frequency, and resistance in the machine; and (iii) equations for finding the current, voltage, frequency, rotor angle, rotor angle delta, angular velocity delta, and maximum power values for each machine affecting instability.

The simulation results of the steady-state stability phenomenon with data on initial assumption values in the form of the emergence of a number of phenomena, namely (a) changes in rotor angle values when achieving stable, (b) achieving time when the rotor delta reaches 0° (stable), (c) achieving time when the delta omega value reaches 0 rad/s (stable), and (d) achieving time when the frequency value reaches 60 hertz (stable). The generating machine with the largest power generation (in this case, machine #3) takes longer to achieve a state of stability again after a change in load. The phenomena in simulating power changes in the installed load on each bus according to the research objectives, namely (a) the active power in machine #2 and machine #3 are always fixed, but the reactive power changes for voltage regulation, so that the generator voltage is equal to the system voltage; (b) the rotor angle value is directly proportional to the change in load power, if the power load rises, the angle value rises and vice versa, so that the condition affects the change in mechanical power, which is when the mechanical power stabilises according to the load power condition, where the mechanical power affects the rotor angle value; (c) the achievement of stable time of rotor angle difference (delta rotor) and stable time of machine #2 frequency is not affected by changes in load power; (d) changes in load power result in achieving a stable time difference in rotor angular velocity (delta omega) relatively slower; (e) generators with large capacities need slower stable achievement times, in which case the capacity on machine #1 is smaller than machine #2 and smaller than machine #3. The damping ratio is inversely proportional to the value of the load power; if the load power rises, the damping ratio decreases; and (f) the system returns to stability when the latter reaches a stable time, i.e. in terms of similarity of frequency, phase sequence, and phase angle. In this simulation, after the machine 3 reaches a stable time.

Completing the conclusions, recommendations are submitted, namely in the implementation of simulations for the analogy of the phenomenon of steady-state stability; it is still necessary to develop by analogy the structure of a larger and more complex electric power system, also through the provision of conditions of simultaneous change on each bus.

Author details

Arief Goeritno[1*], Tapip Hendrawan[2] and Irham Muhammad Azama[3]

1 Electrical Engineering Study Program, Bogor Ibn Khaldun University, City of Bogor, Indonesia

2 Instrumentation and Automation Laboratory, Bogor Ibn Khaldun University, City of Bogor, Indonesia

3 Management Study Program, Muhammadiyah Purwokerto University, Banyumas Regency, Indonesia

*Address all correspondence to: arief.goeritno@ft.uika-bogor.ac.id

IntechOpen

© 2024 The Author(s). Licensee IntechOpen. This chapter is distributed under the terms of the Creative Commons Attribution License (http://creativecommons.org/licenses/by/3.0), which permits unrestricted use, distribution, and reproduction in any medium, provided the original work is properly cited.

References

[1] Kiusalaas J. Numerical Methods in Engineering with MATLAB. 3rd ed. New York: Cambridge University Press; 2015. p. 430. DOI: 10.1017/CBO9781316341599

[2] Moore H. MATLAB for Engineers. 3rd ed. Upper Saddle River: Pearson Education; 2012. p. 650

[3] McMahon D. MATLAB Demystified: A Self-Teaching Guide. New York: McGraw-Hill; 2007. p. 337. DOI: 10.1036/0071485511

[4] IEEE. IEEE standard dictionary of electrical and electronics terms. IEEE Transactions on Power Apparatus and Systems. 1980;**PAS-99(6)**:37a-37a. DOI: 10.1109/tpas.1980.319816

[5] Fuchs EF, Masoum MAS. Power Quality in Power Systems and Electrical Machines. 2nd ed. Salt Lake City: Elsevier; 2015. p. 1115. DOI: 10.1016/C2013-0-18758-2

[6] Hatziargyriou N, Milanovic J, Rahmann C, Ajjarapu V, Canizares C, Erlich I, et al. Definition and classification of power system stability – Revisited & Extended. IEEE Transactions on Power Systems. 2021;**36**(4):3271-3281. DOI: 10.1109/tpwrs.2020.3041774

[7] Kundur P. Power System Stability and Control. New York: McGraw-Hill; 1994. pp. 3-16

[8] Saadat H. Power System Analysis. New York: McGraw-Hil; 1999. pp. 460-526

[9] Anderson PM, Fouad AA. Power System Control and Stability. Hoboken: Wiley-IEEE Press; 2003. pp. 3-12

[10] Murthy PSR. Power System Analysis. Hyderabad: BS Publication; 2007. pp. 259-289

[11] Uppal SL. Electrical Power: Generation, Transmission, Distribution, Witchgear and Protection, Utilization of Electrical Energy and Electric Traction. Delhi: Khanna Publisher; 1984. pp. 654-699

[12] Das D. Electrical Power Systems. New Delhi: New Age International Limited; 2006. pp. 276-306

[13] Ravindranath B, Chander M. Power System Protection and Switchgear. New Delhi: Wiley Eastern Limited; 1977. pp. 1-14

[14] Bakshi UA, Bakshi MV. Protection and Switchgear. Pune: Technical Publication; 2009. pp. 1-34

[15] Ram B, Vishwakarma DN. Power System Protection and Switchgear. 2nd ed. New Delhi: Tata McGraw-Hill Education Private Limited; 2011. pp. 1-31

[16] Goeritno A, Saidah S. Simulation of single-phase to ground fault to anticipate against the sympathetic tripping phenomena. In: Proceeding of the 1st ICETIA 2014; 4th December 2014. Surakarta: UMS; 2014. pp. 229-233. Available from: https://publikasiilmiah.ums.ac.id/xmlui/bitstream/handle/11617/4994/37-Arief%20Goeritno.pdf?sequence=1

[17] Bahrman MP, Johnson BK. Transmission lines: Steady state operation. In: Glover JD, Sarma MS, Overbye TJ, editors. Power System Analysis and Design. 5th ed. Stamford: Cengage Learning; 2012. pp. 233-293

[18] IEEE Committee Report. Proposed terms and definitions for power system stability. IEEE Transaction on Power Apparatus and System. 1982;**101**(7):1894-1898

[19] Savulescu SC. Fast computation of the steady-state stability limit. In: Savulescu SC, editor. Real-Time Stability in Power Systems: Techniques for Early Detection of the Risk of Blackout. 2nd ed. Cham: Springer International Publishing; 2014. pp. 27-61

[20] Goeritno A, Rasiman S. Performance of Bulk Oil Circuit Breaker (BOCB) Influenced by Its Parameters (Case Study at the Substation of Bogor Baru). In: Proceeding of the 3rd International Conference on Engineering, Technology and Industrial Application (ICETIA, 2016); 7th-8th December 2016; Surakarta. Melville: AIP Publishing; 2017. p. 020001. DOI: 10.1063/1.4985446

[21] Goeritno A. Preliminary evaluation for the performance of circuit breaker mediated by SF6 gas. JEEE. 2020;**13**(1): 35-38. Available from: https://electroinf.uoradea.ro/images/articles/CERCETARE/Reviste/JEEE/JEEE_V13_N1_MAY_2020/07%20paper%201115%20GOERITNO.pdf

[22] Yong-Hak K, Tae-Kyun K, No-Hong K, Jin-Boo C, Joon-Young J, Seok-Ha S. Determination of synchronous machine and excitation systems parameters from field tests. IFAC Power Plants and Power Systems Control. 2003;**2003**:757-760. DOI: 10.1016/S1474-6670(17)34562-7

[23] Grainger JJ, Stevenson WD. Power System Analysis. New York, NY: McGraw-Hill, Inc.; 1994. pp. 695-747

[24] Goeritno A. Implementation of the coordination equation for determining the transport-related losses in economic dispatch phenomena. MMEP. 2020;7(2):293-298. DOI: 10.18280/mmep.070216

[25] Ang WK. Energy Function for Power Stability Assessment. Brisbane: The University of Queensland; 2003

[26] Wu M, Lu DD-C. Active stabilization methods of electric power systems with constant power loads: A review. Journal of Modern Power System and Clean Energy. 2014;**2**:233-243. DOI: 10.1007/s40565-014-0066-y

[27] Daware K. Variable loads on power system. 2016. Available from: https://www.electricaleasy.com/2016/01/variable-loads-on-power-system.html

[28] Hassan YF, Merzah AN, Abdul Hassan SK, Tuaimah FM. Optimal reduction of electrical loads based on priorities developed using a genetic algorithm. In: Proceeding of 4th International Conference on Engineering Sciences (ICES 2020); 5th–6th December 2020. Vol. 1067. Kerbala: Materials Science and Engineering; 2021. p. 12127. DOI: 10.1088/1757-899X/1067/1/012127

[29] Sarasua JI, Martinez-Lucas G, Perez-Diaz JI, Fernandez-Munoz D. Alternative operating modes to reduce the load shedding in the power system of El Hierro Island. International Journal of Electrical Power & Energy Systems. 2021;**128**:106755. DOI: 10.1016/j.ijepes.2020.106755

[30] Rajbhandari Y, Marahatta A, Shrestha A, Gachhadar A, Thapa A, Gonzalez-Longatt F, et al. Load prioritization technique to guarantee the continuous electric supply for essential loads in rural microgrids. International Journal of Electrical Power & Energy Systems. 2022;**134**:107398. DOI: 10.1016/j.ijepes.2021.107398

Chapter 3

Power Quality Improvement in Distribution System

Bhavya Kanchanapalli and Sujatha Banka

Abstract

Without electricity, our society would not function in this advanced period. Power supplies of high quality and dependability improve the amount of power that can be transported and distributed. One of the key factors in the power system is power quality. It has grown in significance, especially with the advent of cutting-edge technology. The quality of the power source affects these modern gadgets more strongly. Voltage, current, and frequency at the rated value are referred to as power quality. Any deviation from the usual rating in these amounts is seen as a sign of a power quality issue. It is preferred to employ specialized power devices at the distribution side to get around these power quality issues. DVR and D-STATCOM are chosen among other custom power devices for usage in boosting power quality. The D-STATCOM and DVR are compared in this article using MATLAB software.

Keywords: power quality (PQ), custom power devices (CPD), dynamic voltage restorer (DVR), distribution flexible AC transmission system (D-FACTS), distribution static compensator (DSTATCOM)

1. Introduction

Concerns pertaining to power quality (PQ) are now the most prevalent [1, 2]. The nature of electric loads has completely changed as a result of the extensive use of electronic equipment, including information technology equipment, power electronics, adjustable speed drives (VSD), programmable logic controllers (PLC), and energy-efficient lighting [3]. These loads are both the main contributors to and the main targets of issues with power quality. All of these loads disrupt the voltage waveform because they are non-linear.

Periodically, transient overvoltages (surges/spikes), swells, flickers, uncontrolled voltages, voltage dips/sags, interruptions, and waveform disturbances (power factor, harmonics, among others) can all affect electrical equipment (**Figure 1**) [4].

The word "Power Quality" refers to the quality of voltage, power, and electric current. Power quality concerns such as voltage sags and swells must be handled in the current day. Voltage sags may occur instantaneously and persist from half a cycle to 1 minute. The average voltage will fall by 0.1 to 0.9 pu. Voltage swell, on the other hand, is defined as a rise in rms voltage or current at the power frequency over periods ranging from 0.5 cycles to 1 minute. Magnitudes usually vary between 1.1 and 1.8 pu [5]. As a result, they are a common source of power electronics equipment failure

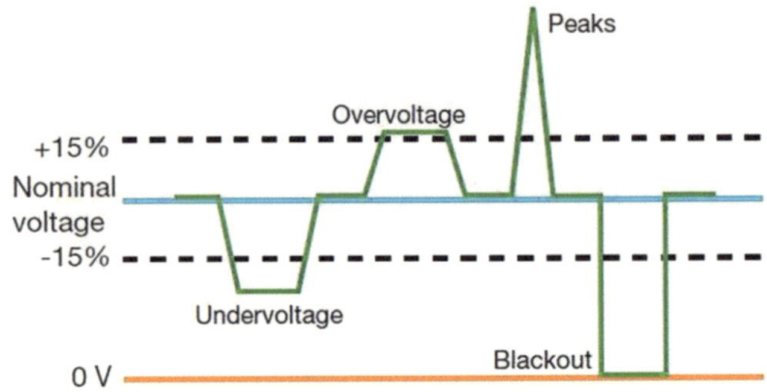

Figure 1.
Various power quality issues in power systems representation.

in contemporary power systems. Voltage swells are less relevant than voltage sags in distribution networks due to their rarity. Voltage sags may occur in both single-phase and three-phase systems. Single-phase voltage sags, on the other hand, have been proven to be prevalent and frequent in the power industry.

2. Faults in the power systems

2.1 Single line to ground faults

In comparison to other fault categories, this problem is less severe and occurs more frequently. There is a 65–70% likelihood of a defect occurring. Any time a line or conductor touches the neutral conductor or falls to the ground, the SLG fault occurs. Due to the imbalanced fault, the equipment is damaged and the system's conditioning is not balanced [2, 4].

2.2 Double line to ground (DLG) fault

DLG faults account for about 15–20% of all defects. Whenever two lines short circuit and come into contact with the earth, a fault is created. In contrast to SLG faults, this fault causes the system to become imbalanced and results in more significant damage to the system whenever it happens [2, 4].

2.3 Line-to-Line (LL) fault

Within the symmetrical faults, this is an unusual fault. Additionally, there is only a 5–10% possibility of this problem occurring. Line-to-line faults occur when two lines short circuit. Within the short-circuited wires, current is flowing due to this problem. When compared to the double line-toto-ground fault, this defect is more severe [2, 4].

Very serious and infrequently occurring faults are symmetrical faults. These flaws, which are also known as balanced faults, are as follows:

2.4 Three-phase to ground fault

Three-phase faults have a 2–3% probability of happening. Because the system maintains its balanced state even in the presence of faults. However, the system's equipment is seriously damaged. When heavy loads are connected and disconnected suddenly, when the grid power supply is unstable, when there are interruptions in the long-distance power supply from the distribution transformer, or when there is a need to boost the power supply to a remote location. The load in a three-phase system is said to be unbalanced for a variety of reasons. Voltage sag, various fault types, and voltage swell are the major causes of voltage disturbance in all of these scenarios. The system may experience a brownout or blackout if the issues are not resolved within a reasonable amount of time. Therefore, short circuit defects in the power system, and voltage swell are therefore regarded as power quality issues in this study [5].

3. Definitions of Power Quality issues

Power Quality refers to the consistency and reliability of electrical power in terms of its voltage, frequency, and other characteristics. Poor power quality can lead to various problems in electrical systems, including equipment malfunction, downtime, and increased maintenance costs. Here are some key definitions related to power quality:

Voltage Sag (or Dip): A short-term decrease in the RMS voltage that lasts from half a cycle to several seconds. Sags can cause sensitive equipment to malfunction or shut down (**Figure 2**) [6].

Voltage Swell: A short-term increase in the RMS voltage. Like sags, swells can negatively affect sensitive equipment (**Figures 3** and **4**) [6].

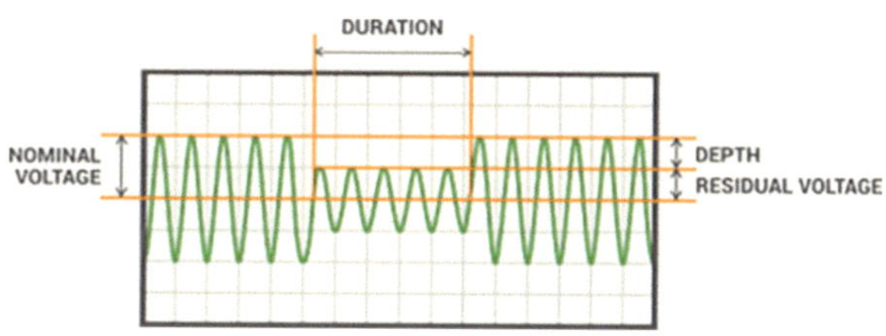

Figure 2.
Voltage sag representation.

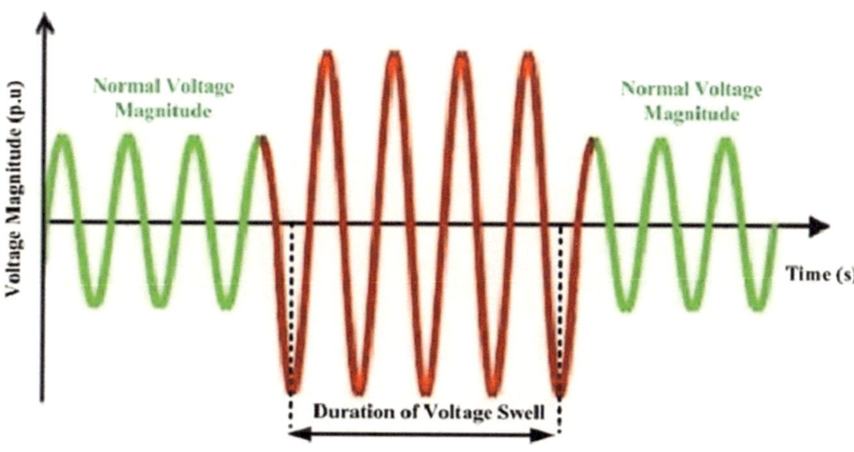

Figure 3.
Voltage swell representation.

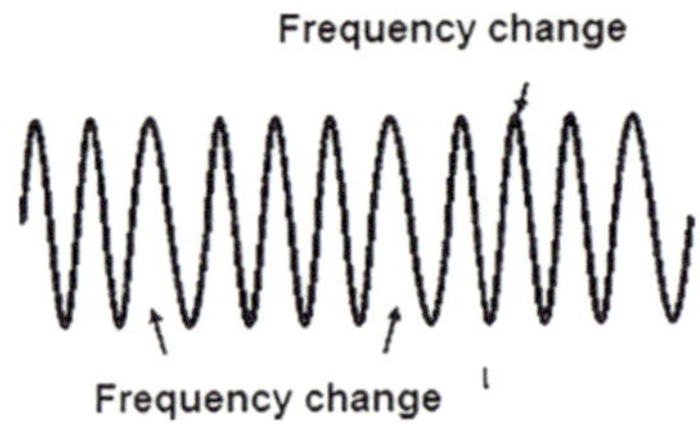

Figure 4.
Frequency variation.

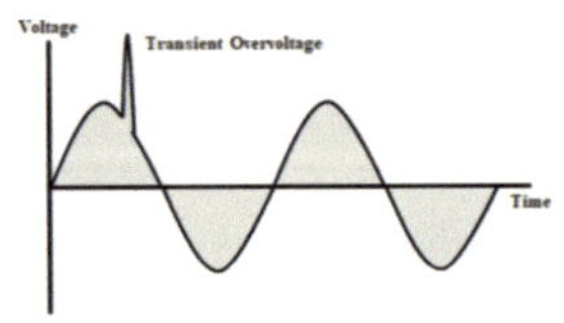

Figure 5.
Trasient over voltage.

Frequency Variations: Deviations from the standard power system frequency (e.g., 50 Hz or 60 Hz). Significant frequency variations can affect the operation of time-sensitive equipment [6].

Transient Overvoltage: A sudden increase in voltage above the normal level, typically lasting for a short duration. Transient overvoltages can damage equipment if not adequately protected (**Figure 5**) [6].

4. Power interruptions

An electrical load losing power is known as a "power interruption," and it is the most basic kind of power quality issue. The various types of power outages are classified based on how long they last.

A momentary disruption occurs when there is a total absence of voltage on one or more phase wires for a duration ranging from 0.5 cycles to 3 seconds.

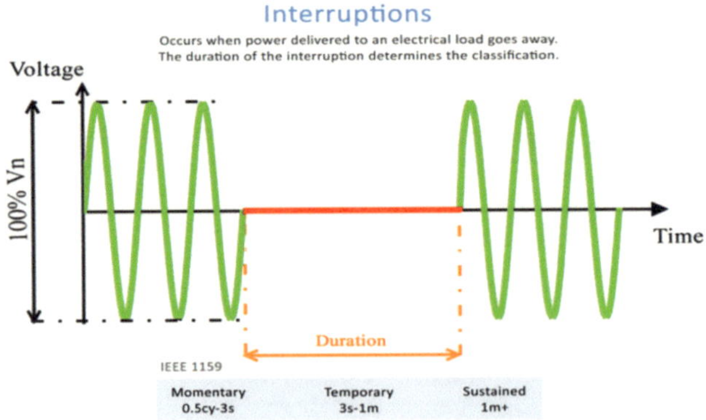

Figure 6.
Various power interruptions.

A momentary disruption occurs when there is a total absence of voltage on one or more phase wires for a duration ranging from 3 seconds to 1 minute.

A total loss of voltage on one or more phase wires for longer than a minute is referred to as a persistent interruption (**Figure 6**) [6].

5. Flicker, transients, and noise

The visual phenomenon known as "flicker" in lighting circuits is the result of repeated voltage decreases [6]. The name "flicker" does not necessarily apply to generic voltage fluctuations; rather, it describes a particularly specific issue with how humans perceive the light generated by incandescent light bulbs.

Arc welders, electric boilers, industrial motors, lasers, photocopying machines, sawmills, and X-ray equipment are a few typical sources of flicker (**Figure 7**).

Transients are caused by spikes superimposed over sine waves of voltage or current, with amplitudes varying from a few volts to several thousand volts. Electronic devices, VFDs, and switching inductive loads usually create low-energy transients continually, but lighting and utility switching usually cause high-energy impulsive transients of brief duration.

Impulsive transients can have a duration of 50 ns to more than 1 ms. The duration of oscillatory transients varies between 0.3 milliseconds and 5 microseconds.

Unwanted, high-frequency oscillations superimposed on a sine wave of alternating voltage or current are referred to as noise. This phenomenon can cause electrical equipment like computers and programmable controllers to malfunction and is typically exacerbated by incorrect grounding.

6. Power factor, unbalance, and harmonics

Electrical loads frequently consist of more than just pure resistance; in an AC system, the impedance is the result of the combination of reactance and resistance. There are two types of reactance: capacitive and inductive, neither of which adds to the power system's "useful" work (**Figure 8**).

The power factor is a metric used to describe the amount of electrical power used to run machines, heat, or lights. A low power factor indicates that a significant quantity of energy is wasted as heat in the system, which often translates into increased energy costs and equipment deterioration.

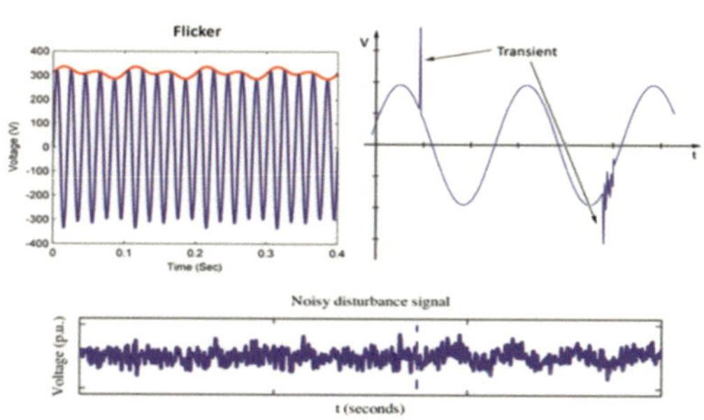

Figure 7.
Flicker, transients, and noise representation.

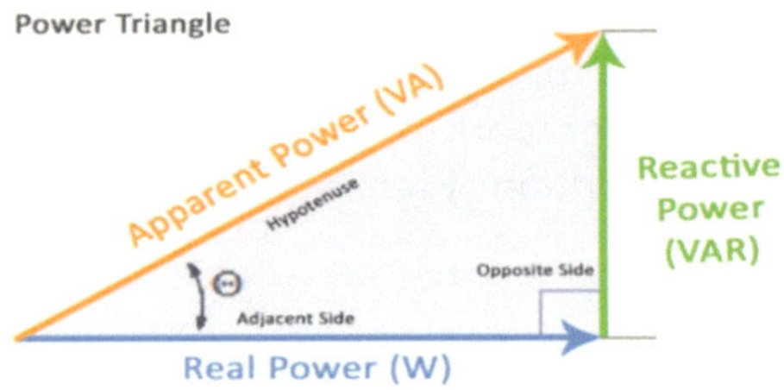

Figure 8.
Power triangle.

Impedances in motors, solenoids, and pumps are usually a mixture of resistance and inductive reactance, and they change depending on the machine's mechanical load. The impedance of a greater capacitive reactance component and a normally tiny resistance. When up the impedance of a capacitor.

The sine waves of the voltage and current in an AC system will move out of phase when reactance is present. The two cancel each other out: voltage leads current with capacitive reactance component and current leads voltage with inductive reactance.

Industrial facilities that have a lot of motors or other inductive loads often have a low power factor. Large commercial and industrial clients are usually charged more by utility providers for poor power factor.

When single-phase loads, such as office equipment and lights, do not consume the same amount of current on each phase, the neutral conductor is put under more stress, which leads to an imbalance in three-phase power systems. Even if the currents may not be in phase with the voltages, an ideal situation arises when the loads are balanced, which means that the voltage and current phases are precisely 120 degrees away from each other (**Figure 9**).

Waveform distortion, known as harmonics, may be found in circuits that include semiconductor-based devices, including test equipment, computers, robots, LED lights, electronic ballasts, and switching power supplies. Higher frequency sine waves on the system by these "non-linear" loads, increasing power loss via lost heat [6].

The excessive heat production may be harmful to a power system. Transformers are particularly vulnerable to harmonic damage because of stray "eddy currents" that circulate inside the iron core of the transformer and generate excessive heat (**Figure 10**).

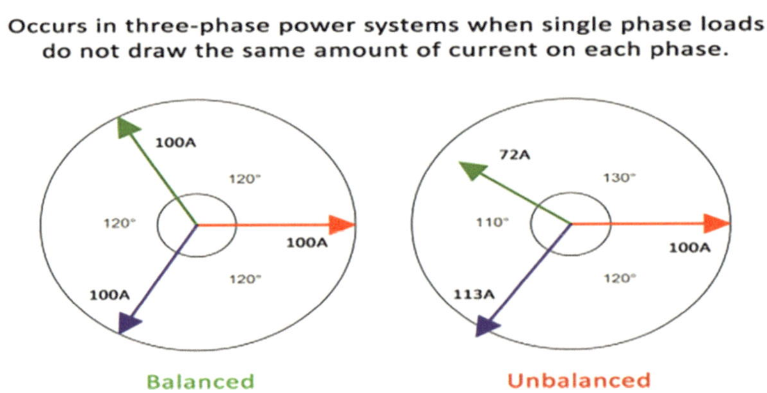

Figure 9.
Unbalance representation.

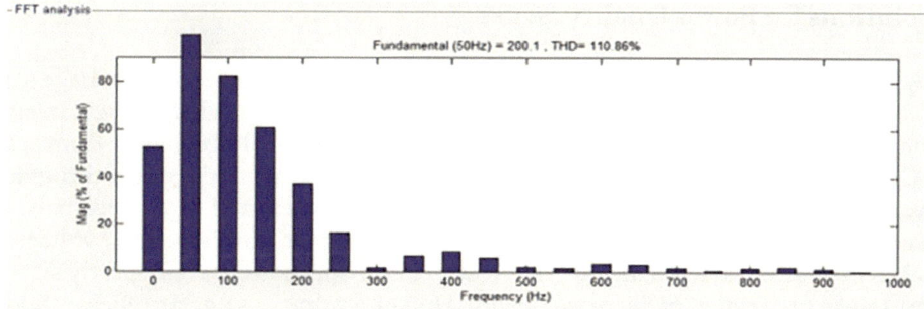

Figure 10.
Harmonic analysis.

The frequency in multiples of the "fundamental" or primary frequency (60 Hz in the US) is used to identify harmonics. In a system operating at 60 Hz, for instance, the third harmonic would be 180 Hz (60 × 3 = 180) and the fifth harmonic would be 300 Hz (60 × 5 = 300) [6].

Power quality metres may be used to measure each harmonic frequency's magnitude; the results are often shown as a harmonic spectrum. When using power quality metres, total harmonic distortion (THD) and total demand distortion (TDD) are sometimes used to reduce harmonic distortion to a single value rather than a spectrum.

7. Power Quality standards

There are a number of industry standards available that address the correct procedures and methods for performing a power quality analysis. These standards should be reviewed to help better understand the science behind monitoring and correcting power quality:

1. ANSI C84.1 - American National Standard for Electric Power Systems and Equipment—Voltage Ratings (60 Hz)

2. IEC 61000 - IEC standards on Electromagnetic compatibility

3. IEEE 519 - IEEE Recommended Practice and Requirements for Harmonic Control in Electric Power Systems

4. IEEE 1159 - IEEE Recommended Practice for Monitoring Electric Power Quality

5. IEEE 1250 - IEEE Guide for Identifying and Improving Voltage Quality in Power Systems

6. IEEE 1668 - IEEE Trial-Use Recommended Practice for Voltage Sag and Short Interruption Ride-Through Testing for End-Use Electrical Equipment Rated Less than 1000 V

7. IEEE 1789 - IEEE Recommended Practices for Modulating Current in High-Brightness LEDs for Mitigating Health Risks to Viewers

8. Solutions for Power Quality issues

A Custom Power Device (CPD) is one efficient technique to decrease voltage sags and swells caused by power quality concerns [7]. DSTATCOM and DVR are currently among the most common power devices due to their use of Voltage Source Converter (VSC) technology. Because these critical powers are delivered to the client through power electronics controller devices, they are sometimes referred to as specialized power devices. Some research has given promising concepts for decreasing voltage sag induced by motor startup using D-STATCOM. These solutions are only partly successful since they can only decrease the duration of a voltage sag occurrence or boost the voltage level during it. They are also not adaptable solutions since, according to research that employed DVR to reduce voltage sag induced by starting an induction motor, the efficacy of compensation might be reduced when motor rating changes. The DVR operates by injecting an in-phase series voltage into the load together with the incoming supply, just enough to return the voltage to its pre-sag condition. It has a proven track record of avoiding voltage sags in practical installations. **Figure 11** shows various Power Quality Issues.

On several occasions, efforts have been undertaken to provide an active and flexible solution for power quality problems. One of these power quality methods that is widely used to lower harmonics is lossless passive filters with L-C tuned components. Passive filters are preferred due to their low starting cost and high initial efficiency. It has a number of drawbacks, including utility impedance, fixed compensation, resonance with supply and loads, and instability. These limitations have been circumvented by the use of active power filters. Active power filters can be made in shunt, series, or hybrid forms. Hybrid is created by combining series and

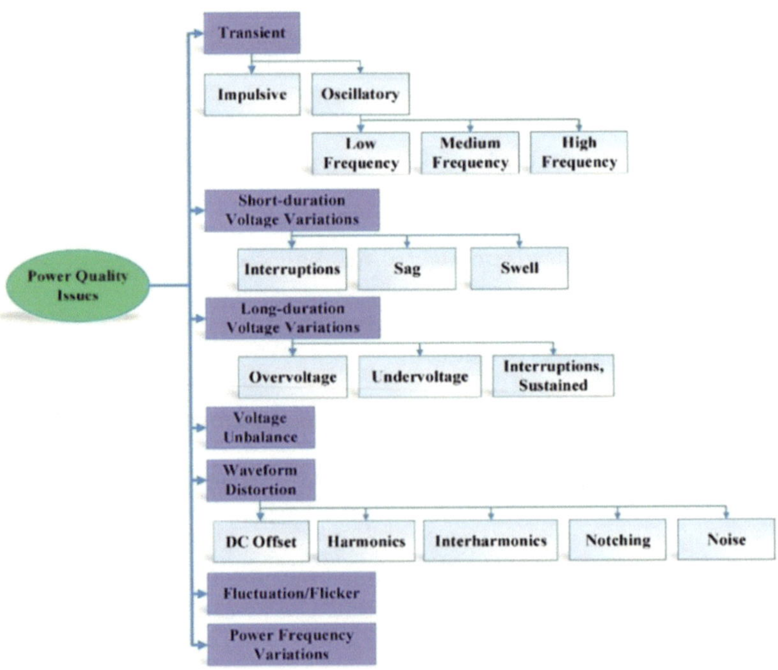

Figure 11.
Power quality issues in distribution system.

shunt types. Voltage-based distortions, whereas corrected using series APF corrects current-based distortions, whereas corrected using shunt APF. Hybrid APF is used to filter high-order harmonics. In typical applications, their rating might occasionally be fairly near to load (up to 80%), which is an issue. This makes it impossible to achieve the appropriate level of power quality. Power outages and dissatisfied consumers are the outcomes of this. Innovative power electronics controller devices have been made available during the past few decades in an effort to increase the reliability of the distribution system and deal with power disturbance concerns. The creation of power electronics controller devices led to the birth of customized power.

The "Custom Power" strategy intends to primarily satisfy the requirements of industrial and commercial customers. In order to provide the degree of quality that picky customers want, "custom power" refers to tools for integrating power electronics controllers into a power distribution system.

They are also known as customized power devices since these significant powers are delivered to users through these power electronics controller devices. Most of them are available for sale as goods and perform effectively at medium distribution levels. VSI is often used to develop specialized power devices since it can maintain the DC bus voltage on its own utilizing a large DC capacitor. The two major categories of bespoke power equipment are:

1. Type 1: Network reconfiguration

2. Type 2: Compensating (**Figure 12**)

8.1 D-STATCOM

One of the Shunt connected DFACTS devices is the Distributed Static Compensator (DSTATCOM). The DSTATCOM system is made up of three major components: a Voltage Source Converter (VSC), a collection of coupling reactors, and a controller. A DSTATCOM placed in a power system generates a controlled alternating current voltage source through a voltage source inverter (VSI) coupled to a DC

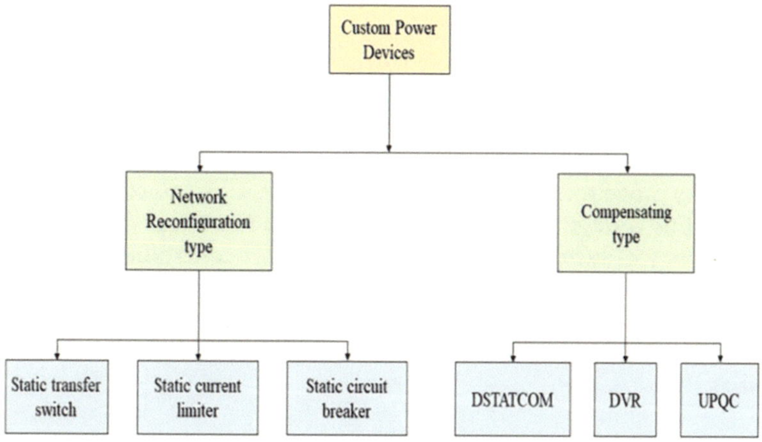

Figure 12.
Types of custom power devices.

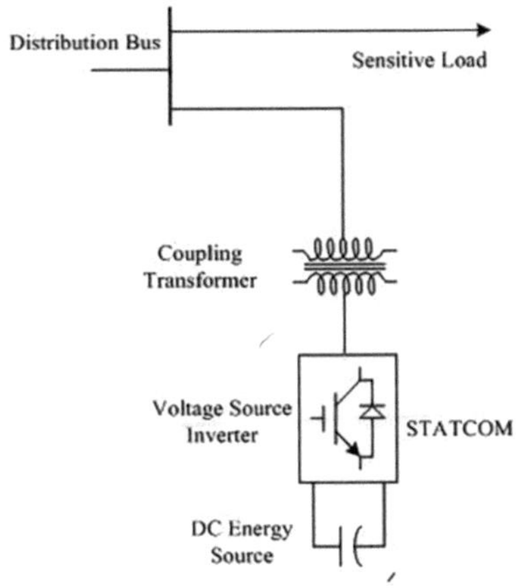

Figure 13.
Symbol of DSTATCOM.

capacitor (energy storage device). In general, the alternating current voltage source appears behind a transformer leakage reactance. The voltage differential across this reactance causes active and reactive power transfer between the power supply and the DSTATCOM. The DSTATCOM is linked to the power networks if there is a voltage-quality issue. All necessary voltages and currents are monitored and supplied into the controller, where they are compared to the orders. The controller then conducts feedback control and generates a series of switching signals to operate the power converter's primary semiconductor switches (IGBTs, which are employed at the distribution level) (**Figure 13**).

8.2 The operation of DSTATCOM

The firing angle is used to adjust the firing angle. The output voltage of the VSI should ideally be in phase with the voltage of the bus to which the DSTATCOM is attached. The dc side capacitance is kept constant at a given voltage in steady state, and there is no practical power exchange except for losses. The DSTATCOM differs from other reactive power producing devices (such as shunt capacitors, static VAR compensators, and so on) in that energy storage is optional and only necessary for system imbalance or harmonic absorption. The DSTATCOM has two control goals in place. The first is the alternating current voltage control of the power supply at the bus to which the DSTATCOM is attached. The other is DC voltage control across the DSTATCOM's capacitor. Shunt reactive power injection is well recognized for its ability to adjust bus voltage. Two voltage regulators are developed for these functions in a traditional control system. AC voltage regulator for controlling bus voltage and DC voltage regulator for controlling capacitor voltage [2, 4]. Both regulators in the simplest scheme are proportional integral (PI) controllers.

8.2.1 Operating modes

It is operated in three modes.

Mode 1: If the DSTATCOM voltage is greater than the Grid Voltage, DSTATCOM supplies the Reactive Power to the Grid, so this mode is named as Capacitive Mode.

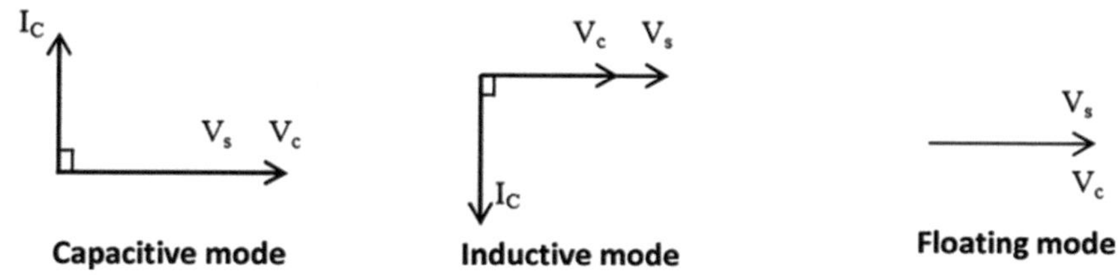

Figure 14.
Modes of operation.

Mode 2: If the Grid voltage is greater than the DSTATCOM voltage, the Reactive Power is absorbed by the DSTATCOM, so this mode is called Inductive Mode.

Mode 3: If the grid voltage is equal to DSTATCOM Voltage, then there is no exchange of Reactive Power. So this is called as Floating Mode (**Figure 14**).

A Distribution Static Synchronous Compensator (DSTATCOM) offers several advantages in electrical distribution systems, especially in addressing power quality issues. Here are some of the key advantages of using DSTATCOM:

1. Reactive Power Compensation: DSTATCOM can provide dynamic reactive power support to maintain system voltage levels within desired limits, thereby improving voltage stability and ensuring efficient power transfer.

2. Voltage Regulation: By injecting or absorbing reactive power as needed, DSTATCOM helps regulate voltage levels, reducing voltage fluctuations and ensuring consistent voltage supply to critical loads.

3. Harmonics Mitigation: DSTATCOM can mitigate harmonics in the distribution system by injecting counteracting currents, thereby improving the quality of power and reducing the risk of equipment damage and inefficiencies caused by harmonics.

4. Power Factor Correction: DSTATCOM can correct power factor issues by supplying or absorbing reactive power, helping utilities and industrial facilities comply with regulatory requirements and improve overall system efficiency.

5. Fast Response Time: DSTATCOM offers rapid response capabilities, allowing it to quickly compensate for voltage variations, harmonics, and other power quality issues, thereby minimizing disruptions and ensuring reliable operation of electrical systems.

6. Flexible Installation: DSTATCOM can be installed in various configurations (shunt, series, or combined) based on specific application requirements, allowing for flexible integration into existing distribution systems.

7. Energy Savings: By improving power quality, reducing losses associated with poor power factor, and optimizing voltage levels, DSTATCOM can help reduce energy consumption and improve overall system efficiency.

8. Enhanced System Reliability: By addressing voltage fluctuations, harmonics, and other power quality issues, DSTATCOM helps enhance system reliability, reduce downtime, and extend the lifespan of electrical equipment and components.

9. Scalability: DSTATCOM systems can be designed and scaled to meet specific application requirements, making them suitable for a wide range of distribution system configurations and load profiles.

10. Compliance with Standards: DSTATCOM solutions can help utilities, industrial facilities, and commercial establishments comply with various power quality standards, regulations, and guidelines related to voltage regulation, harmonics mitigation, and power factor correction.

In summary, DSTATCOM offers a range of benefits for improving power quality, enhancing system reliability, and ensuring efficient operation of electrical distribution systems. By addressing voltage regulation, reactive power compensation, harmonics mitigation, and other power quality issues, DSTATCOM helps utilities and end-users maintain high-quality power supply, optimize energy efficiency, and reduce operational costs.

8.3 Dynamic voltage restorer (DVR)

In 1996, a 12.47 kV system in Anderson, South Carolina became the first DVR to be deployed in North America. DVRs have since been used to safeguard crucial loads in the utility, semiconductor, and food processing industries. One of the most efficient PQ devices for resolving voltage sag issues nowadays is the dynamic voltage restorer. However, cost and installation limitations have restricted its use to situations where a reliable power source is obviously needed.

DVR is a static var. device that has found use in a number of transmission and distribution systems. It is a series compensation device that protects sensitive electric loads from power quality issues such as voltage sags, swells, imbalance, and distortion using voltage source converters (VSC) (**Figure 15**).

8.3.1 Principle of operation

The Dynamic Voltage Restorer's is to inject a voltage of the necessary size and frequency to enable it to restore the load side voltage to the proper amplitude and waveform, even when the source voltage is imbalanced or distorted [2, 4].

In general, it uses a pulse width modulated (PWM) inverter structure and Gate Turn Off Thyristor (GTO) solid-state power electronic switches. At the load side, the DVR has the ability to independently produce or absorb actual and reactive power. In other words, the DVR is built of a solid-state DC to AC switching power converter

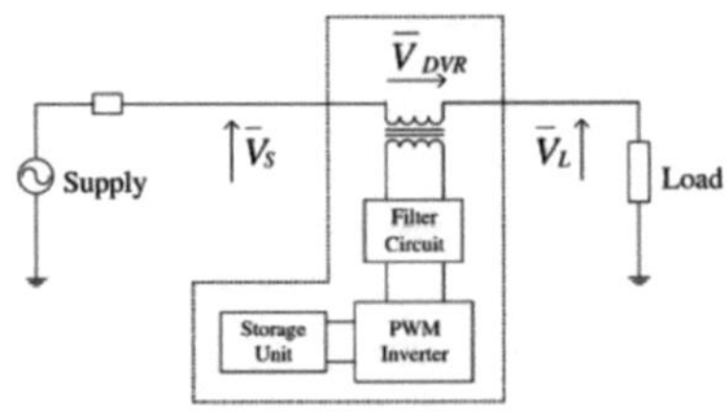

Figure 15.
Symbol of DVR.

that injects a sequence of three-phase AC output voltages into the transmission and distribution lines. The commutation mechanism for reactive power demand and an energy source for actual power need are the sources of the injected voltage.

Depending on the DVR's manufacturer and design, the energy source may change. DC capacitors, batteries, and energy extracted from the line via a rectifier are a few types of energy sources that have been used.

The DVR system's practical injection voltage capacity is 50% of nominal voltage. Due to this, DVRs may successfully defend against sags of up to 50% for periods of time as little as 0.1 seconds. Furthermore, the majority of voltage sags seldom ever go below 50%.

The destructive consequences of voltage swells, voltage imbalance, and other waveform distortions are also lessened by using the Dynamic Voltage Restorer.

There are different methods to control the DVR. They are

i. Distinguish the occurrence of sag or swell in the system.

ii. Compute the offset voltage.

iii. Pulse output of the PWM inverter and stopping it when the problem is resolved.

8.3.2 MATLAB/Simulink model of DSTATCOM

Let us consider a test system where a sudden RL-load is applied all of a sudden and the response of the output voltage is analyzed using the Simulink platform [8].

Due to the sudden application of the load, voltage sag is observed, and this should be compensated by using our Custom Power Devices to transfer constant voltage to the load (**Figures 16–21**).

8.4 MATLAB/Simulink design of DVR

A Dynamic Voltage Restorer (DVR) is a power electronic device primarily designed to mitigate voltage sags and interruptions in electrical distribution systems [8]. Here are some of the key advantages of using a DVR (**Figures 19–21**):

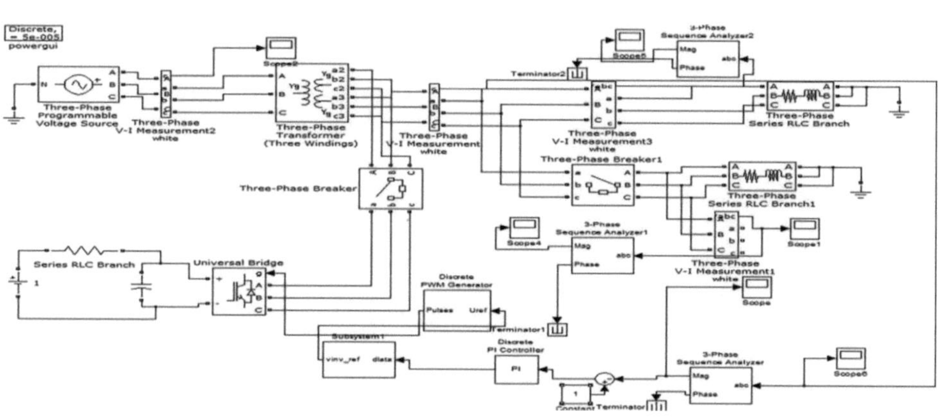

Figure 16.
Simulation model of the test system with DSTATCOM in MATLAB/Simulink for RL load.

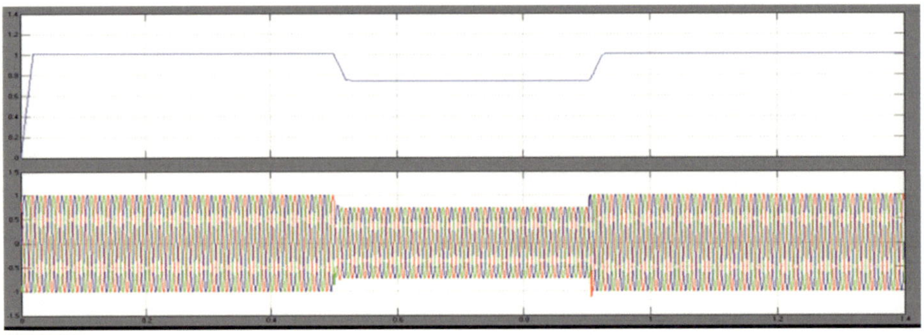

Figure 17.
Voltage sag at the load side.

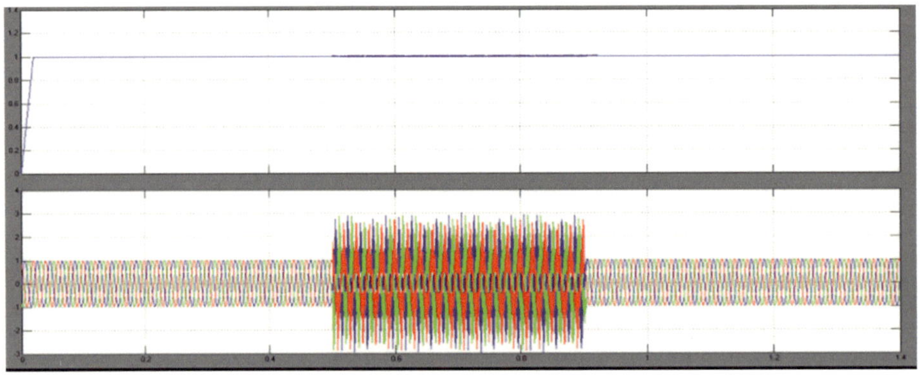

Figure 18.
Compensated voltage through the introduction of DSTATCOM on the load side.

1. Voltage sag mitigation: One of the primary advantages of a DVR is its ability to rapidly inject voltage to compensate for voltage sags, thereby restoring the voltage waveform and ensuring uninterrupted operation of sensitive equipment and processes [1].

2. Fast Response Time: DVRs offer very fast response times, typically within a few milliseconds, to detect voltage disturbances and inject compensating voltage, minimizing downtime and disruptions for critical loads [1].

3. Improved Power Quality: By mitigating voltage sags and interruptions, DVRs help improve overall power quality, reduce equipment downtime, and enhance the reliability of electrical distribution systems [1].

4. Protection for Sensitive Equipment: DVRs provide essential protection for sensitive electronic equipment, such as computers, servers, industrial machinery, and other critical loads, by ensuring a stable and consistent power supply.

5. Flexible Installation: DVRs can be installed at critical points in the distribution system to protect specific loads or areas prone to voltage sags, allowing for targeted mitigation and optimization of power quality.

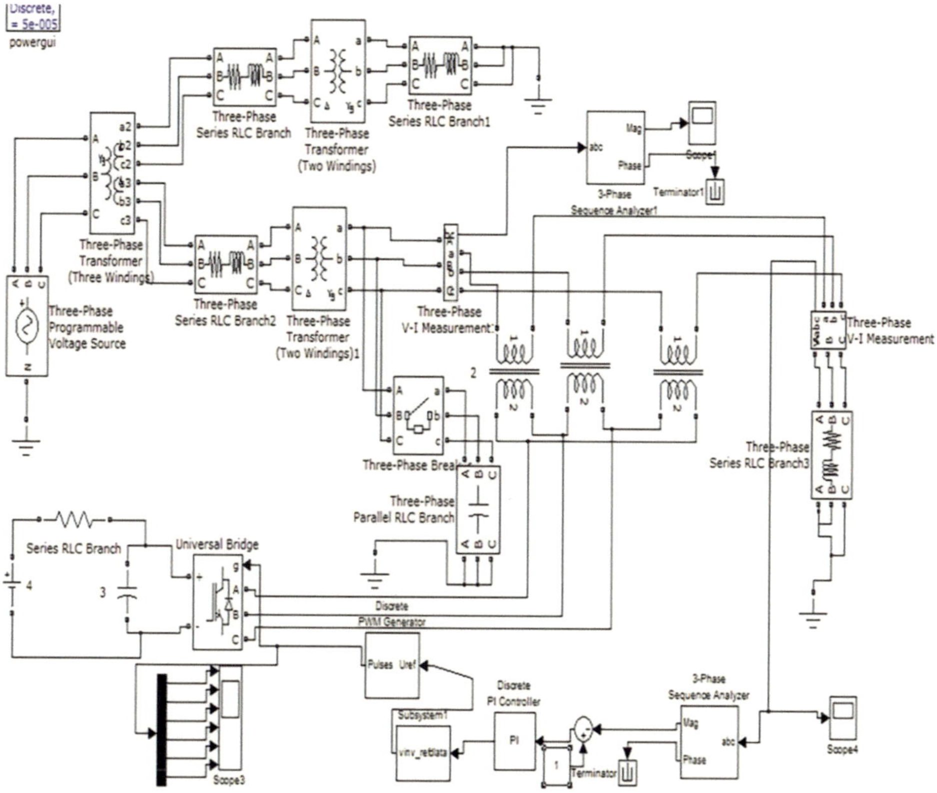

Figure 19.
Simulation model of the test system with DVR in MATLAB/SIMULINK for RL-load.

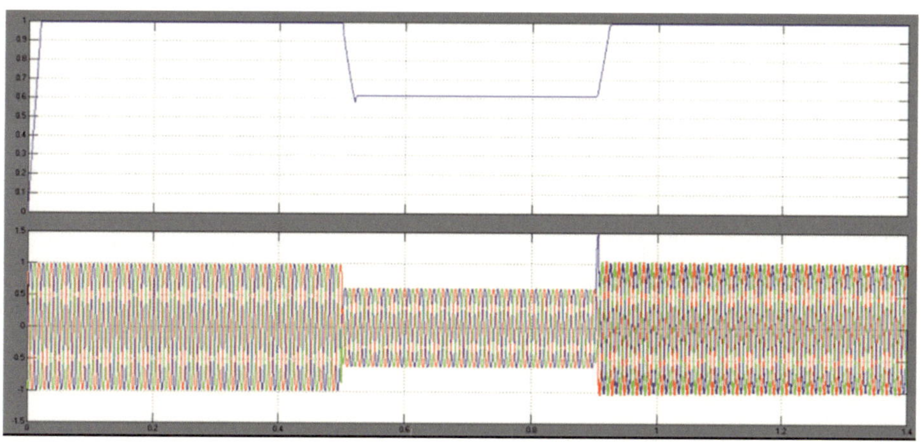

Figure 20.
Voltage at the load point without DVR in MATLAB/SIMULINK for 3-phase fault.

6. Compatibility with Renewable Energy Sources: DVRs can be integrated with renewable energy sources, such as solar and wind power systems, to ensure stable and reliable power supply, especially in areas with fluctuating renewable energy output.

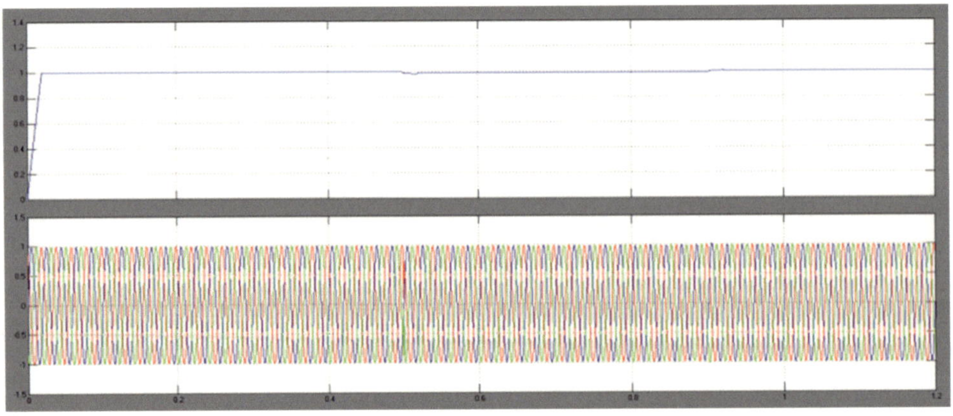

Figure 21.
Voltage at the load point with DVR in MATLAB/SIMULINK.

7. Cost Savings: By reducing equipment downtime, preventing production losses, and protecting sensitive equipment from voltage sags and interruptions, DVRs can lead to significant cost savings for industrial facilities, commercial establishments, and utilities [1].

8. Extended Equipment Lifespan: DVRs help extend the lifespan of electrical and electronic equipment by providing a stable and consistent power supply, reducing wear and tear, and minimizing the risk of damage from voltage disturbances [1].

9. Compliance with Standards: DVRs help utilities, industrial facilities, and commercial establishments comply with various power quality standards, regulations, and guidelines related to voltage sag mitigation, ensuring adherence to industry best practices and regulatory requirements [1].

10. Scalability and Modularity: DVRs can be designed and configured to meet specific application requirements, making them scalable and modular solutions for addressing voltage sags and interruptions in various distribution system configurations and load profiles [1].

In summary, DVRs offer a range of benefits for improving power quality, protecting sensitive equipment, reducing downtime, and ensuring reliable operation of electrical distribution systems. By providing fast and effective voltage sag mitigation, DVRs help utilities and end-users maintain high-quality power supply, optimize equipment performance, and minimize operational risks and costs.

8.5 Comparison of DVR and DSTATCOM

Dynamic Voltage Restorer (DVR) and Distributed Static Compensator (DSTATCOM), two different devices used in power systems to improve voltage stability and power quality. Although both are designed to mitigate voltage disturbances, they operate in different ways and have distinct features. Let us compare them based on their characteristics (**Table 1**) [9].

Parameter	DVR	DSTATCOM
Operation	A DVR is a power electronics-based device that is connected in series with the distribution line. It continuously monitors the voltage at its point of connection and injects a voltage in series with the line to compensate for voltage sags or swells.	A DSTATCOM is also a power electronics-based device, but it is connected in parallel with the distribution line. It actively injects reactive power into the system to stabilize voltage, compensate for reactive power demand, and improve power factor.
Function	The primary function of a DVR is to mitigate voltage sags and swells caused by various disturbances such as faults, sudden load changes, or system switching operations.	The main function of a DSTATCOM is to regulate voltage and maintain power factor at the point of connection, compensating for reactive power demand and improving voltage stability.
Compensation Time	DVRs are generally capable of providing rapid voltage compensation within a few milliseconds, as they are designed to quickly respond to voltage disturbances.	DSTATCOMs can also respond quickly to voltage fluctuations and provide compensation, but their response time might be slightly longer compared to DVRs due to their different configuration.
Installation Location	DVRs are typically installed close to critical loads to protect them from voltage variations and ensure continuous operation.	DSTATCOMs are often installed at distribution substations or near industrial loads to enhance voltage stability and power quality for the entire distribution network.
Application Scope	The DVR is specifically designed to address voltage quality issues such as voltage sags and swells in sensitive loads like data centers, industrial processes, and critical facilities.	DSTATCOMs are more versatile and can be applied to various scenarios, including voltage regulation in distribution systems, power factor correction, and reactive power compensation for individual loads or groups of loads.
Cost-effective	DVRs can be more expensive due to their high-speed power electronics and sophisticated control systems.	DSTATCOMs are generally more cost-effective and easier to implement, making them suitable for broader applications.

Table 1.
Comparison of DVR and DSTATCOM.

9. Conclusion

The Power Quality issues, including voltage dips, swells, and interruptions, as well as mitigation strategies for specialized power electronic devices like DVR and D-STATCOM, have been covered in this chapter. Comprehensive findings are provided together with the design and implementations of DVR and DSTATCOM for voltage sags, interruptions, and swells. In order to operate the electronic valves in the VSI utilized in the DSTATCOM and DVR, a new PWM-based control system has been created. Unlike basic frequency switching strategies that are already included in MATLAB/Simulink, this PWM control system requires just voltage data. It is perfect for low-voltage bespoke power applications because of this feature. The simulation results of the simulations demonstrated that the DVR had comparatively superior voltage regulating capabilities [4]. It was also noted that the rating of the DC storage device affects DSTATCOM's ability to regulate voltage and provide power compensation. The simulation results shown exhibit high accuracy.

Author details

Bhavya Kanchanapalli* and Sujatha Banka
BVRIT HYDERABAD College of Engineering for Women, Hyderabad, India

*Address all correspondence to: bhavya.k@bvrithyderabad.edu.in

IntechOpen

© 2024 The Author(s). Licensee IntechOpen. This chapter is distributed under the terms of the Creative Commons Attribution License (http://creativecommons.org/licenses/by/3.0), which permits unrestricted use, distribution, and reproduction in any medium, provided the original work is properly cited.

References

[1] Vishwakarma J, Sharma AK. Simulation and comparison of DVR and D-STATCOM used for voltage sag mitigation at distribution side. International Journal of Recent Research in Electrical and Electronics Engineering (IJRREEE). 2015;**2**(2):197-205

[2] Parmar PC, Kathiria MK, Sonagra KA. Operation of DVR and D-STATCOM for load voltage control in distribution side. International Journal for Innovative Research in Multidisciplinary Field. 2017;**3**(4):159-165. ISSN – 2455-0620

[3] Bapaiah P. Power quality improvement by using DSTATCOM. International Journal of Emerging Trends in Eelctrical and Electronics. Apr 2013;**2**(4):1-12

[4] Ghosh H, Kumar Saha P, Kumar Panda G. Performance comparison between DVR and DSTATCOM used for load voltage control in distribution side. In: 2012 International Conference on Advances in Power Conversion and Energy Technologies (APCET), Mylavaram, India. 2012. pp. 1-6. DOI: 10.1109/APCET.2012.6302026

[5] Thirupathaiah M, Venkata Prasad P, Ganesh V. Performance analysis of DVR and DSTATCOM in distribution network for voltage sag and swell conditions. Power Research-A Journal of CPRI. 2016;**12**:723-730

[6] Rojin RK. A review of power quality problems and solutions in electrical power system. International Journal of Advanced Research in Electrical, Electronics, and Instrumentation Engineering. Nov 2013;**2**(11):5605-5614. ISO 3297: 2007

[7] Singh A, Surjan BS. Comparative study of facts devices for mitigation of voltage sag-swell in distribution system. Indian Journal of Science and Technology. 2016;**9**(34):1-8

[8] Dey S. Comparison of DVR and D-Statcom for voltage quality improvement. International Journal of Emerging Technology and Advanced Engineering. 2014;**4**(10):187-193

[9] Hossain E, Tür MR, Padmanaban S, Ay S, Khan I. Analysis and mitigation of power quality issues In distributed generation systems using custom power devices. IEEE Access. 2018;**6**:16816-16833. DOI: 10.1109/Access.2018

Chapter 4

Harmonics Emissions and Power System Grid Resilience in Electric Vehicle Charging

Makawi Diab Hraiz and Juan Andrés Martín García

Abstract

This chapter focuses on the harmonics emissions and power system grid resilience in electric vehicle (EV) charging. With the increasing demand for EVs, understanding the effects of harmonics on grid stability is critical. Converters in EV charging operations can contribute to high harmonics levels, causing disturbances and stressing the power system. The scope of the chapter encompasses challenges related to harmonics and emphasizes the need to ensure grid resilience. It also highlights the importance of understanding the correlations between harmonics emissions, operational parameters, and the increasing adoption of electric vehicle charging. This chapter also aims to discuss and present various harmonic mitigation techniques that can enhance power quality in electric vehicle integration.

Keywords: electric vehicle charging, harmonics emissions, power system grid resilience, voltage distortion, fast charging

1. Introduction

The surge in carbon dioxide emissions and volatile oil prices have opened doors for alternative technologies like electric vehicles (EVs). By harnessing sustainable green energy, the electrification of transportation presents a viable solution to address the goal of reducing greenhouse gas emissions. Electric vehicles offer the advantage of zero emissions and high energy conversion efficiency, setting them apart from conventional combustion engines. However, the rapid integration of EVs can pose a significant challenge to existing electric distribution systems, compounded by the current insufficiency of EV charging infrastructure.

The widespread adoption of electric vehicles globally has considerable momentum as a key solution to addressing environmental issues and reducing fuel dependency in the transport sector [1]. In 2022, Europe witnessed a significant growth in EV registrations, reaching a total of 1,729,000 vehicles, marking a 7% increase compared to the previous year [2]. European governments have implemented a range of incentives to support and accelerate the expansion of EVs, aiming for high levels of penetration. For instance, Spain has set an ambitious target of reaching 4 million electric vehicles by 2030 [3], which will contribute to a substantial surge in EV adoption in the coming years. The growing demand market for electric vehicles as a more environmentally friendly transportation option has raised concerns about the impact of charging these vehicles on electric grids.

In the near future, it is anticipated that the penetration of EVs into low-voltage residential networks will rise significantly. Due to its ease and the lack of charging infrastructure, overnight home charging is anticipated to be preferred by EV buyers. Due to the high penetration rates of EV charging stations and the growing number of EVs, the power infrastructure is currently facing enormous challenges. Consequently, a technical analysis of how EV charging affects the distribution system is necessary. EVs are becoming more and more popular every year, and they will soon supplant the current transportation system. To effectively plan for and accommodate this growing EV load while maintaining power quality, detailed technical studies of the impact of EV charging on the distribution system become imperative [4, 5]. These plans and research should include a precise evaluation of the EV load and its effects on power quality indices to facilitate efficient infrastructure planning. Manufacturers, consumers, and researchers all share a keen interest in the development of electric vehicles and their relationship with grid infrastructure. The integration of electric cars into distribution networks introduces several power quality concerns, with harmonic emissions being a notable issue. These challenges become particularly pronounced in scenarios with a high percentage of electric vehicle penetration, taking into account the clustering effect during the charging process [6].

The current capacity of electrical distribution systems is not projected to sufficiently accommodate such a rapid influx of EV penetration. Moreover, the integration of EVs into the distribution network has adverse effects on power quality due to the harmonics generated by these vehicles. In situations where loads exhibit non-linear behavior, drawing current in non-sinusoidal and distorted waveforms while connected to the network, has a detrimental impact on public networks designed to operate with sinusoidal voltage at rated power. These non-linear loads result in the creation of harmonics, leading to voltage distortion for customers, increased mechanical stress, confusion among protection equipment, and decreased efficiency of system components. EVs introduce harmonics into electrical systems through their non-linear charging behavior. Despite efforts to improve power electronics circuits and power factor correction, the harmonics generated by EV chargers are expected to be significant due to the high charging currents. Therefore, addressing harmonic emissions is crucial for maintaining power quality in the context of increasing EV adoption.

2. The influence of electric vehicles on harmonic emissions

The influence of harmonics in EV charging systems is a significant consideration due to its potential effects on power quality and the overall stability of the electrical network. In this context, multiples of the AC power supply's fundamental frequency are referred to as harmonics. EV chargers have the potential to produce harmonics with unwanted frequencies that vary from the typical sinusoidal waveform.

The problem of power quality in EV charging systems is one of the main effects of harmonics. Increased distortion from harmonic currents and voltages can cause voltage waveform distortion and can cause electrical equipment to overheat. The effectiveness of the power distribution network may be affected and other linked devices may be impacted by this distortion. Furthermore, harmonics may increase the system's energy losses and lower its overall energy efficiency.

Effective mitigation requires an understanding of the specifics of how harmonics are created in EV charging systems. Harmonics are produced in part by the non-linear design of EV chargers, especially those that use power electronics for rectification and conversion procedures. Waveform distortion results from the sudden pulses of current drawn by these nonlinear loads. Retaliators and converters used in the charging process cause abrupt changes in current, which reintroduces harmonic currents into the grid. The charging EV process involves rectifiers and converters that introduce rapid changes in current, leading to harmonic currents being injected back into the electrical network.

Most studies emphasized the significance of the issue of harmonics, especially with traditional chargers. High harmonics could lead to decreased lifespans of distribution network components, such as transformers and cables [7]. These harmonics can have several consequences. Firstly, they can lead to voltage distortion within the electrical system, affecting the quality of power supplied to other devices and appliances connected to the grid. This distortion can cause disruptions and reduce efficiency in the operation of various electrical equipment. Secondly, the presence of high-frequency harmonics can result in increased losses within transformers and cables. This is primarily because these harmonics cause a non-uniform distribution of current and subsequently lead to additional heating in these components. Such losses can impact the overall efficiency of the electrical system. Lastly, the integration of EVs into the grid, with the associated harmonics, poses various challenges. These challenges include managing harmonic distortion levels, addressing imbalances in the system, ensuring proper voltage regulation, and mitigating potential losses in transformers. Addressing these issues is essential to maintain the reliability and stability of the electrical grid in the face of increasing EV adoption. Harmonics are introduced into the system by non-linear equipment, such as EV chargers, as illustrated in **Figure 1**.

Two basic measures have been developed to estimate the amount of harmonic distortion and develop mitigation solutions. Total harmonic distortion (THD), which measures the percentage of the Root Mean Square (RMS) of the harmonic frequency components concerning the fundamental frequency component for both voltage V_1 and current I_1, is the first of these measures. Here, as shown in Eqs. (1) and (2), V_h and I_h stand for the RMS values of various-order harmonics in voltage and current [9].

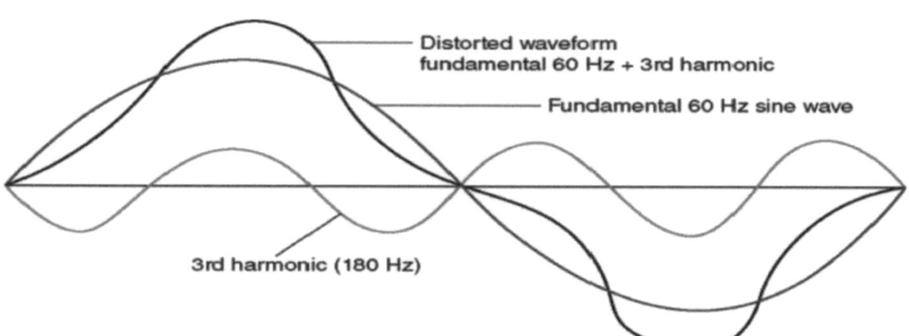

Figure 1.
Harmonic distortion in which the waveform is made up of fundamental and 3rd harmonic [8].

$$V^{THD} = \frac{\sqrt{V_2^2 + V_3^2 + \ldots + V_h^2}}{V_1} \tag{1}$$

$$I^{THD} = \frac{\sqrt{I_2^2 + I_3^2 + \ldots + I_h^2}}{I_1} \tag{2}$$

Total Demand Distortion (TDD) is the second measure for evaluating harmonic distortion [10]. The following formula is used to calculate TDD, where I_{ref} stands for the nominal current:

$$TDD = \frac{\sqrt{\sum_{h=2}^{n} I_h^2}}{I_{ref}} \tag{3}$$

International standards prescribe specific thresholds for current harmonics. For example, the IEC 61000 standard establishes a 13% limit for THD, while the IEEE 519 standard specifies a 5% threshold for TDD. These standards serve as regulatory benchmarks to ensure that levels of harmonic distortion in electrical systems remain within acceptable ranges.

2.1 Voltage distortion

Harmonics can distort the voltage in the electrical system, affecting the quality of power supplied to other devices and appliances connected to the grid.

Simultaneous fast charging of multiple electric vehicles will result in voltage distortion surpassing permissible limits. As per the EN 50160 standard, the total harmonic distortion of the supply voltage, encompassing all harmonics up to the 40th, should remain below 8% [11]. The substantial current demands of fast charging significantly influence voltage quality, particularly at the connection point, emphasizing its critical role in ensuring voltage stability.

The modeling of the fundamental and h^{th} harmonic currents from the nonlinear load situated at bus i, which has real power P and reactive power Q, is expressed as follows:

$$I_i^1 = \left[(P_i + jQ_i)/V_i^1\right] \tag{4}$$

In Eq. (5), R(h) represents the ratio of the hth harmonic current to its fundamental counterpart

$$I_i^1 = R(h)\ I_i^1 \tag{5}$$

The harmonic voltages are determined by solving the following load-flow equation and consequently [8], the voltage at bus i is defined as shown in Eq. (7)

$$Y^h\ V^h = I^h \tag{6}$$

$$|V_i^h| = \left(\sum_{h=1}^{H} |V_i^h| 2\right)^{1/2} \tag{7}$$

2.2 Transformer and cable losses

The presence of high-frequency harmonics can lead to increased losses in transformers and cables. This is because these harmonics cause non-uniform current distribution and result in additional heating.

One of the challenges associated with EV battery charging comes from the potentially high harmonic currents associated with the conversion of AC power system voltages to DC EV battery voltages. Improvements in EV battery technology and government incentives will potentially contribute to higher EV penetration. Despite the advantages of this green technology, battery charging current harmonics injected into the grid incite an increase in transformer losses, followed by its temperature rise and lifetime reduction. The performance and ability of distribution transformers suffer under non-sinusoidal load operation, resulting in a lower power factor and quality [12].

The level of current harmonics circulating in a transformer winding can affect its operating temperature and lifetime. Although the existing standards mainly consider the impact of harmonics up to 2 kHz, higher frequency harmonics generated by high-power converters utilized in renewable energy sources can also contribute to the temperature rise of a transformer. Pulse Width Modulated voltage generated by power converters can generate significant high-frequency harmonics at its switching frequency. The switching frequency of converters in high-power applications is mainly between 2 and 9 kHz [13].

The transformer load P_L including current harmonics is presented in P_{EC} as the eddy-current loss, and P_S is the other stray loss. The harmonic loss factor, F_L, is applied to the winding eddy loss to indicate the heating due to harmonic load current and determine the transformer capabilities when supplying power to a load.

The harmonic loss factor for other stray losses, F_{HS} is employed to represent the heating impact as a result of other stray losses in liquid-filled transformers I^2 is the rms current [14].

$$P_L = I^2(1 + P_{EC}\,F_L + P_S\,F_{HS}) \tag{8}$$

Table 1 summarizes the various studies and their findings related to the impact of current harmonics on transformers, particularly in the context of EV chargers and nonlinear loads.

It's noteworthy that underground cables are more susceptible to the impacts of EV-related harmonics compared to overhead transmission lines. Consequently, this area necessitates more stringent requirements and standards for charging stations. During the charging process, the maximum values of harmonics in cables consistently exceed those in overhead transmission lines. This discrepancy primarily arises from the lower surge impedance of cables compared to overhead lines, attributed to the closer proximity of conductors, which increases capacitance and reduces inductance [20].

2.3 Battery conditions

In order to carefully manage the energy transfer between various powertrain subsystems and guarantee the delivery of necessary torque and power to the vehicle's wheels, EV powertrains heavily rely on power electronic components and electric machinery. Notably, along the powertrain's direct current bus, these power subsystems have the potential to produce unwanted electrical harmonics. As a result, in

Electrical systems and transformers studies	Harmonics findings	Impact on transformer
Transformer losses and heat [15]	Current harmonics induce heat, causing thermal stress. 50% of transformer loss of life attributed to nonlinear loads.	Reduced lifespan due to heat stress.
Grid-connected inverters [16]	Experiment: Inverters generate low-order current harmonics (<2 kHz) leading to a 1.2°C insulation temperature increases at the rated load.	8.3% reduction in expected lifespan at rated load.
EV charger harmonics [17]	Assessment of EV chargers: harmonic loss factor predicts a significant potential decrease in transformer lifespan if the EV charger load exceeds 80% of the rated load.	Potential lifespan reduction with high EV charger load.
Industrial system load currents [18]	Simulation results: with approximately 80% nonlinear load, the transformer aging acceleration factor reaches about 2.97, indicating a significant reduction in its lifespan.	Notable lifespan reduction with high nonlinear load.
EV home chargers and power quality [8]	Study of EV home chargers: quadratic relation found between Total Harmonic Distortion (THDi) and transformer life consumption. Recommended THDi threshold of 25–30% for acceptable increase in life consumption.	THDi impacts transformer lifespan.
EV charger harmonics and distribution capacity [19]	Study on distribution system capacity: results indicate the 10 kV cable overload at 27.25% EV charger penetration due to harmonics, compared to 30.74% without harmonics. Suggested solution involves adding filters to EV chargers.	Distribution capacity and overload management.

Table 1.
Effects of harmonics on transformer performance and electrical system.

addition to the basic DC current, the battery of the EV may be subjected to unfavorable oscillations that appear as both high- and low-frequency current variations. On the high voltage DC bus of EVs, empirical testing under real-world circumstances has revealed significant current harmonic disturbances ranging from 50 Hz to 4 kHz. Comprehensive studies have been conducted to understand the effects of these harmonics on the degeneration of battery systems within the small body of available literature [21].

3. Enhancing power system grid resilience

The overall relationship between power systems and grids takes center stage when examining the integration of electric vehicles. The power system, includes the generation, transmission, and distribution of electricity, while the power grid specifically refers to the interconnected equipment facilitating this energy transfer. The dynamic interaction between these components becomes crucial as EV adoption picks up speed. Aligning the grid with the changing EV landscape requires careful station location, smart charging algorithms, and the incorporation of renewable energy sources. Maintaining power quality and grid stability simultaneously requires tackling harmonic issues brought about by nonlinear loads from EV chargers. In order to create an integrated framework that guarantees the smooth integration of EVs into current power infrastructures, utilities, regulators, and technology developers must cooperate together.

Harmonic losses in the grid system affect energy efficiency because they raise resistive losses, which lower the total power factor. Higher generation capacity may be required to meet demand if there is a decline in the effective power provided to the loads as a result of a reduced power factor. As a result, the grid infrastructure is strained and might need to be upgraded to handle the increasing demand for EV charging.

Ensuring the reliability and dependability of energy infrastructure is critical, particularly concerning electric vehicles, requiring for enhancing the energy system grid's resilience. In the era of electric vehicles, improving grid resilience is a complex problem that needs a variety of technological, governmental, and policy solutions. These techniques can be used to build a more enduring and efficient power supply that can support the increasing number of electric vehicles while enduring interruptions and disturbances.

To achieve this, it is crucial to take into account a variety of network reliability techniques and strategies, the most crucial of which are:

1. Developing the current grid infrastructure: electrical network modernization by integrating many advanced technologies such as sensors, smart meters, and automation to monitor and control the network more effectively. This allows for faster response times during outages and improve load management, even as demand for electric vehicle charging increases. One of the pivotal aspects of strengthening the electrical grid infrastructure is enhancing the cybersecurity of the grid in order to protect against cyber threats that may disrupt the energy system.

2. Solutions for power outages: ensuring continuous electric vehicle charging during power outages requires the implementation of three strategies. Firstly, the adoption of distributed energy resources, such as photovoltaic systems and small-scale wind turbines, enables local energy generation and storage. Secondly, by establishing small grids within communities, especially in areas exposed to severe weather events, to generate, distribute and consume energy locally. Electric vehicles can play a role in these microgrids by acting as mobile energy storage units and energy sources. Thirdly, equipping charging stations with backup power sources like batteries or generators, and integrating this with a grid resilience strategy. These strategies ensure uninterrupted EV charging and reduces reliance on the main grid during power outages, thus enhancing overall resilience.

3. Government regulations and public awareness: grid resilience should be prioritized while regulatory frameworks should support the seamless integration of electric vehicles into the grid. This can be done by encouraging investments in cutting-edge fast-charging station technologies and grid modernization. To reduce grid stress, regulated and responsible charging behaviors must be encouraged. Public awareness about the importance of proper charging behavior during grid stress or power outages. Mobile applications that provide information on grid stress times may be created by local authorities to help educate electric vehicle owners.

Moreover, when considering the grid's size, the harmonics add complexity to the process of figuring out the right capacity and layout. Harmonic currents have the potential to overload transformers and conductors, which might affect sizing calculations that are typically based on linear load assumptions. Because harmonics are not sinusoidal, the sizing parameters need to be reconsidered in order to ensure that the grid infrastructure can accommodate the additional demand from EV charges without affecting reliability.

4. Harmonics emissions with charging rate

Harmonic emissions are impacted by a number of charging-related issues for electric vehicles. Various elements, including as charging rates, power distribution, and usage patterns, might be considered operational parameters. The management of harmonics in EVs is aided by the identification of probable correlations and their implications through the analysis of these parameters. For the purpose of creating management or mitigation methods for harmonics, investigating these relationships is crucial. This section delves into a more detailed exploration of the influence of various operational parameters within EVs on harmonics and explains why it's crucial to analyze these parameters for harmonics management.

The impact of charging rates, a critical operational parameter, on harmonics in the context of EVs warrants thorough investigation. The swiftness of the charging process, particularly in fast-charging scenarios, is a subject of notable concern. While fast charging is instrumental in expediting the adoption of EVs, it introduces rapid fluctuations in power demand, which may engender harmonics and voltage variations. Fast charging, with its abrupt power demand shifts, can be a source of harmonic distortions and may lead to substantial voltage perturbations and losses [22].

The incorporation of electric vehicles primarily occurs within residential and commercial settings, and the speed of charging can potentially pose challenges related to harmonic distortion [6]. Several studies investigate both voltage and current harmonics, placing specific emphasis on the fast-charging process, particularly when it occurs within a cluster of chargers interconnected to the same electrical feeder [23]. The study cited in reference [24] observed a total harmonic distortion in the current from fast charging stations, which reached 32%.

When compared to regular or slow charging, fast charging entails much faster charging rates. This indicates that during the charging process, the power usage changes quickly and significantly. The electrical system may be subjected to harmonics as a result of these quick power shifts. While the low EV penetration levels with typical charging rates will result in acceptable harmonic levels [25].

To improve these issues, the adoption of standard charging protocols or the implementation of intelligent charging systems capable of adjusting charging rates in alignment with the grid's capacity have demonstrated efficacy in mitigating harmonics. Moreover, a comprehensive analysis of the influence of these fast-charging stations on both low and high-order harmonics becomes imperative. This stems from the fact that the chargers utilized in these stations possess the potential to introduce harmonics into distribution networks at varying harmonic orders, necessitating a comprehensive analysis of the impact on the grid's harmonics.

5. Managing harmonics and mitigation

Addressing higher harmonics, which are multiples of the fundamental frequency, can be relatively straightforward. By using active or passive filters or by incorporating suitable modulation techniques into the management of power electronics switches, these problems can be solved. These techniques work well to decrease higher harmonics' negative influences on the power system. In contrast, extra complexity is introduced when dealing with lower harmonics, like the third, fifth, seventh, eleventh, and comparable orders. It is a difficult task to filter out these lower harmonics while preserving the integrity of the primary frequency signal. Because of their close closeness within the frequency spectrum, it becomes difficult to differentiate lower harmonics from the fundamental frequency. There are methods for canceling out lower harmonics, but they often come with practical and financial challenges. Harmonic cancellation methods, though available, tend to be neither cost-effective nor straightforward to implement from a technical standpoint [26].

Using an Active Power Filter APF to mitigate the harmonics produced by the EV charging station is a practical approach, given the difficulties in lowering harmonics from an EV charger that depends on passive power factor correction. Because they are more widely used than series APFs, shunt APFs are the better alternative. This is mainly because of their well-established, mature technology and ease of installation [27]. Uncontrolled high-speed charging caused the THDv to increase to 11.4%, above the recommended 8% limit, a solution to the harmonic problem was proposed by using the control of PV inverter as an active filter [28].

In certain research studies, it was proposed that the utilization of a smart charger can lead to a substantial reduction in THD when compared to conventional chargers. For example, in a study referenced as [29], the smart charger was found to draw sinusoidal current and maintain a unity power factor. Typically, ultra-high frequency switching is used by some inverters to lower emissions of harmonic distortion. To improve their chargers, manufacturers could include power factor correction circuits in the rectification stage. These circuits could use filters and pulse width modulation (PWM) techniques to handle harmonic issues [30].

Combining these approaches can result in a robust and effective system for EV chargers that effectively reduces harmonic distortions and ensures the health of the power distribution network as a whole. Additionally, smart grid solutions can further enhance communication between chargers and the grid. This will allow for intelligent charging scheduling. Research and development should be supported through educational programs, standards compliance, and regular system audits to ensure harmonic health and encourage the development of innovative harmonically friendly charging methods.

6. Conclusions

The integration of electric vehicles into the grid can pose challenges related to harmonic distortion levels, imbalances, voltage regulation, and potential losses in transformers. Harmonic distortion in electrical systems is a well-known concern, as it results in waveforms that deviate from the ideal sinusoidal pattern. The increasing use of non-linear power electronic devices and the proliferation of sensitive electronic loads have raised significant issues related to the safety and the reliable operation of electronic equipment. EVs emit harmonics due to their non-linear charging behavior, and these harmonics can impact the quality and efficiency of the electrical grid and connected devices. Managing and mitigating these harmonics is a critical consideration in the widespread adoption of electric vehicles.

Most investigations have shown that using normal EV chargers frequently results in harmonic levels that are undesired. If these high harmonic values are not reduced, it may be damaging to the lifespan of the transformers and cables that make up the distribution network. However, the negative effects of charger-induced harmonics can be greatly reduced by utilizing well-thought-out design practices for EV charger circuits, efficient control strategies, and the integration of filters into the charger circuit. More studies are necessary to address the issues raised by fast charging stations. These efforts ought to concentrate on removing harmonic distortion emissions from chargers through the improvement of the filters that are integrated into these charging systems, specifically for the purpose of preventing low-order and high-order harmonics.

The infrastructure of EV charging stations needs to minimize harmonics strategically on a larger system size. In order to bring harmonic distortion into the permitted range, this includes carefully placing harmonic mitigation devices, such as filters. This proactive strategy is essential for maintaining the electrical grid's stability and dependability, particularly in view of the increasing number of electric vehicles on the road and the harmonic issues they provide.

Acknowledgements

Makawi Diab Hraiz gratefully acknowledges the University of Cádiz for postdoc-toral support with a Margarita-Salas fellowship, funded by the Ministry of Universi-ties of the Government of Spain through the European Recovery Instrument "Next Generation EU", of the European Union (Order UNI/551/2021, of 26 May).

Author details

Makawi Diab Hraiz* and Juan Andrés Martín García
Electrical Engineering Department, University of Cadiz, Cadiz, Spain

*Address all correspondence to: makawidiab.makawihraiz@alum.uca.es

IntechOpen

© 2024 The Author(s). Licensee IntechOpen. This chapter is distributed under the terms of the Creative Commons Attribution License (http://creativecommons.org/licenses/by/3.0), which permits unrestricted use, distribution, and reproduction in any medium, provided the original work is properly cited.

References

[1] Habib S, Khan MM, Abbas F, Ali A, Hashmi K, Shahid MU, et al. Risk evaluation of distribution networks considering residential load forecasting with stochastic modeling of electric vehicles. Energy Technology. 2019;7: 1900191

[2] European Environment Agency. New registrations of electric vehicles. Available from: https://www.eea.europa.eu/ims/new-registrations-ofelectric-vehicles#footer [Accessed: November 29, 2023]

[3] Navas-Anguita Z, García-Gusano D, Iribarren D. Long-term production technology mix of alternative fuels for road transport: a focus on Spain. Energy Conversion and Management. 2020;**226**: 113498

[4] Iqbal MN, Kütt L, Daniel K, Asad B, Shams Ghahfarokhi P. Estimation of harmonic emission of electric vehicles and their impact on low voltage residential network. Sustainability. 2021;**13**(15):8551

[5] Iqbal MN, Kütt L, Lehtonen M, Millar RJ, Püvi V, Rassõlkin A, et al. Travel activity based stochastic modelling of load and charging state of electric vehicles. Sustainability. 2021;**13**(3):1550

[6] Caro LM, Ramos G, Rauma K, Rodriguez DFC, Martinez DM, Rehtanz C. State of charge influence on the harmonic distortion from electric vehicle charging. IEEE Transactions on Industry Applications. 2021;**57**(3): 2077-2088

[7] Nour M, Chaves-Ávila JP, Magdy G, Sánchez-Miralles Á. Review of positive and negative impacts of electric vehicles charging on electric power systems. Energies. 2020;**13**(18):4675

[8] Srivastava A, Manas M, Dubey RK. Electric vehicle integration's impacts on power quality in distribution network and associated mitigation measures: A review. Journal of Engineering and Applied Science. 2023;**70**:1-29

[9] Thomas D, D'Hoop G, Deblecker O, Genikomsakis KN, Ioakimidis CS. An integrated tool for optimal energy scheduling and power quality improvement of a microgrid under multiple demand response schemes. Applied Energy. 2020;**260**:114314

[10] Guo J, Yang J, Ivry P, Serrano C. Analysis of the impacts of V2G chargers on LV grid harmonics. In: 2020 9th International Conference on Renewable Energy Research and Application (ICRERA). IEEE; 2020. pp. 88-92

[11] Wang Q, Zhou N, Wang J, Wei N. Harmonic amplification investigation and calculation of electric vehicle charging stations using three-phase uncontrolled rectification chargers. Electric Power Systems Research. 2015;**123**:174-184

[12] Said DM, Nor KM, Majid MS. Analysis of distribution transformer losses and life expectancy using measured harmonic data. In: 14th International Conference on Harmonics and Quality of Power-ICHQP 2010. IEEE; 2010. pp. 1-5

[13] Yaghoobi J, Alduraibi A, Martin D, Zare F, Eghbal D, Memisevic R. Impact of high-frequency harmonics (0–9 kHz) generated by grid-connected inverters on distribution transformers. International Journal of Electrical Power & Energy Systems. 2020;**122**:106177

[14] EE recommended practice for establishing liquid-filled and drytype

power and distribution transformer capability when supplying nonsinusoidal load currents - Redline. In: IEEE Std C57.110–2008 (Revision of IEEE Std C57.110–1998) - Redline. 2008. pp. 1-67

[15] Resende JW, Samesima MI, At-aujo SCN. Analysis of transformer loss of life driving nonlinear industrial loads by the finite elements approach. In: Industry Applications Conference. 1995. pp. 75-79

[16] Awadallah MA, Xu T, Venkatesh B, Singh BN. On the effects of solar panels on distribution transformers. IEEE Transactions on Power Delivery. 2016; **31**(3):1176-1185

[17] Alame D, Azzouz M, Kar NC. Impact assessment of electric vehicle charging on distribution transformers including state-of-charge. In: 2018 IEEE 61st International Midwest Symposium on Circuits and Systems (MWSCAS). 2018. pp. 607-610

[18] Cazacu E, Petrescu M, Ioniță V, Petrescu L. Nonsinusoidal load current effect on the electrical and thermal operating parameters of oil filled power distribution transformers. In: 2018 18th International Conference on Harmonics and Quality of Power (ICHQP). 2018. pp. 1-6

[19] Zhou C, Wang H, Zhou W, Qian K, Meng S. Determination of maximum level of EV penetration with consideration of EV charging load and harmonic currents. IOP Conference Series: Earth and Environmental Science. 2019;**342**:012010

[20] Amamra SA, Tripathy Y, Barai A, Moore AD, Marco J. Electric vehicle battery performance investigation based on real world current harmonics. Energies. 2020;**13**(2):489

[21] Xu Y, Xu Y, Chen Z, Peng F, Beshir M. Harmonic analysis of electric vehicle loadings on distribution system. In: 2014 IEEE International Conference on Control Science and Systems Engineering. IEEE; 2014. pp. 145-150

[22] Lucas A, Bonavitacola F, Kotsakis E, Fulli G. Grid harmonic impact of multiple electric vehicle fast charging. Electric Power Systems Research. 2015; **127**:13-21

[23] Dharmakeerthi CH, Mithulananthan N, Saha TK. Impact of electric vehicle fast charging on power system voltage stability. International Journal of Electrical Power & Energy Systems. 2014;**57**:241-249

[24] Cheng L, Wu B, Chen H, Ma W, Zhang G. Load forecasting and harmonic analysis for EV charging station. In: Proc. of the 2018 IEEE Innovative Smart Grid Technologies-Asia (ISGT Asia). 2018

[25] Foosnæs AH, Jensen AN, Nordentoft NC. Report: Case Studies of Grid Impacts of Fast Charging, Edison Consortium, D 4.4.3 & D 4.4.4 Copenhagen. 2011. pp. 1-62

[26] Mazin HE, Xu W. Harmonic cancellation characteristics of specially connected transformers. Electrical Power System Research. 2009;**79**(12): 1689-1697. DOI: 10.1016/j.epsr.2009.07.006

[27] Liu Q, Peng L, Kang Y, Tang S, Wu D, Qi Y. A novel design and optimization method of an LCL filter for a shunt active power filter. IEEE Transactions on Industrial Electronics. 2014;**61**:4000

[28] Nguyen VL, Tran-Quoc T, Bacha S. Harmonic distortion mitigation for electric vehicle fast charging systems. In: Proceedings of the 2013 IEEE Grenoble

Conference, France; 16–20 June 2013. IEEE; 2013. pp. 1-6

[29] Gomez J, Morcos M. Impact of EV battery chargers on the power quality of distribution systems. IEEE Transactions on Power Delivery. 2003;**18**:975

[30] Basta B, Morsi WG. Low and high order harmonic distortion in the presence of fast charging stations. International Journal of Electrical Power & Energy Systems. 2021;**126**:106557

Chapter 5

A Review on Utilization of Electric Vehicles for Mitigating the Power Quality Issues in Power Systems

Mohammad Mehdi Amiri, Saleh Aghajan-Eshkevari, Mohammad Ali Rahimi and Ali Samari

Abstract

The widespread adoption of EVs can alleviate strain on power grids and enhance power quality in multiple ways. First, the integration of EVs enables better utilization of renewable energy sources by serving as mobile energy storage units. During periods of peak renewable energy generation, excess power can be stored in EV batteries, reducing curtailment and maximizing resource efficiency. Second, smart charging infrastructure and vehicle-to-grid (V2G) technology allow EVs to interact with the grid intelligently. Through V2G systems, EVs can supply power back to the grid during high-demand periods, effectively functioning as decentralized energy storage units. This bidirectional energy flow helps stabilize voltage and frequency fluctuations, ultimately improving overall power quality. Furthermore, EVs can facilitate load balancing by enabling charging during off-peak hours, spreading the electricity demand more evenly throughout the day. This approach minimizes the strain on the grid during peak times and enhances system stability. In summary, electric vehicles not only reduce emissions and promote sustainability but also contribute to optimizing power generation, storage, and distribution, leading to a more resilient and higher-quality power supply. In this chapter, the impact of electric vehicles on improving the quality of power in the voltage and frequency sections has been investigated.

Keywords: electric vehicles, power quality, power system, frequency control, voltage stability, voltage unbalance

1. Introduction

Electric vehicles (EVs) represent a greener alternative to conventional vehicles as they do not emit CO_2. Auto manufacturers are actively producing EVs and promoting their adoption over traditional vehicles. To charge EVs, charging stations are used. However, the amount of power EVs draw during charging varies due to factors like battery technology and charging methods (slow, fast, ultra-fast/flash charging). Flash charging, for instance, can fully charge an EV in under a minute by drawing a significant amount of power from the grid. The diversity in EV charging behaviors poses challenges to the distribution grid, including overloading, stability issues, power

IntechOpen

quality problems, voltage imbalances affecting other users on the same grid, etc. In the near future, EV adoption is set to surge, with annual sales projected to reach 23–43 million by 2030. This will considerably strain the grid, potentially requiring a capacity increase of 640–1110 TWh by 2030 [1]. The widespread adoption of electric vehicles (EVs) brings both benefits and challenges to the electrical distribution network. The most important challenges are depicted in **Figure 1**.

a. *Overloading*: The sudden and simultaneous charging of multiple EVs, especially during peak periods, can lead to increased demand on the distribution network. If the infrastructure is not designed to handle this increased load, it can result in the overloading of transformers, feeders, and other components, potentially leading to equipment failures and service disruptions [2, 3].

b. *Stability problems*: Rapid changes in power demand caused by EV charging can lead to voltage fluctuations and frequency deviations. These variations can impact the overall stability of the distribution network, potentially causing voltage sags, swells, or even voltage collapses. Voltage instability can result in equipment damage and reduced system reliability.

c. *Power quality problems*: The introduction of EVs can cause power quality issues such as harmonic distortion and poor power factor. These issues can lead to increased losses, decreased efficiency, and potential interference with sensitive equipment connected to the same network.

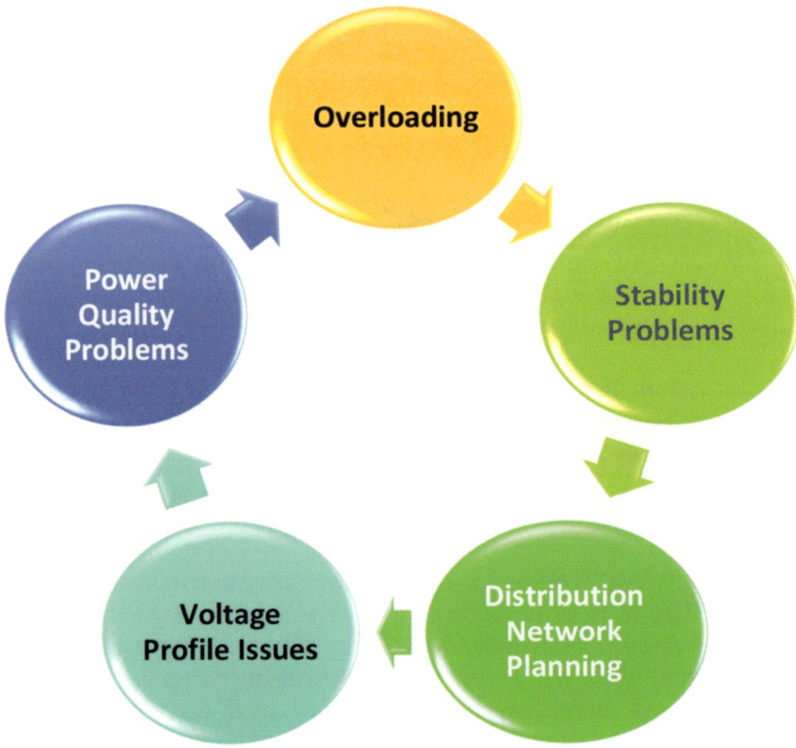

Figure 1.
EVs challenges to the electrical distribution network.

d. *Voltage profile issues*: The dynamic nature of EV charging can lead to voltage fluctuations that affect not only the charging stations but also other connected consumers. Low-voltage conditions can hinder proper charging, while high-voltage conditions can stress and damage equipment. Maintaining a stable voltage profile becomes a challenge when dealing with the variable power demands of EVs.

e. *Distribution network planning*: The increased load due to EV charging might require costly upgrades to the distribution network's infrastructure. Proper network planning and investment are necessary to accommodate the growing number of EVs without compromising the system's reliability [4].

To address these challenges and minimize their adverse impacts, various strategies and solutions can be considered:

- *Smart charging*: Implementing smart charging solutions that consider the distribution network's capacity and load constraints can help distribute the charging load more evenly and reduce the likelihood of overloading [5].

- *Demand response programs*: Incentivizing EV owners to participate in demand response programs, where charging is scheduled during off-peak periods, can help mitigate peak load demands and reduce stress on the distribution network.

- *Grid-integrated charging infrastructure*: Developing charging stations that are integrated with the grid and can communicate with the utility can enable better load management and coordination, thus reducing the impact of EV charging on the network [6].

- *Advanced metering and monitoring*: Using advanced metering and monitoring systems, utilities can gain real-time insights into network conditions and adjust operations accordingly to maintain stability and power quality [7].

- *Distribution network upgrades*: In cases where demand from EVs is substantial, targeted distribution network upgrades may be necessary to accommodate the increased load safely and efficiently.

- *Energy storage integration*: Integrating energy storage systems into the distribution network can help smooth out load fluctuations caused by EV charging and provide additional stability.

- *Regulatory and policy measures*: Governments and regulatory bodies can play a role by implementing policies that incentivize responsible charging behaviors, promote smart charging infrastructure deployment, and ensure that the grid can handle the transition to EVs.

On the other hand, we can also mention the importance of power quality. Power quality is integral to the reliable and efficient operation of power systems and the equipment they serve. Addressing power quality issues requires a combination of monitoring, analysis, and corrective measures to ensure a stable and high-quality power supply for consumers, industries, and critical infrastructure. Here are some key reasons highlighting the importance of power quality in power systems:

a. *Equipment performance and reliability*: Many modern devices and equipment are sensitive to variations in voltage, frequency, and waveform quality. Poor power quality, such as voltage sags, surges, harmonics, and flickers, can lead to equipment malfunctions, downtime, and premature wear and tear. Maintaining good power quality ensures the reliable and optimal performance of a wide range of equipment, from industrial machinery to sensitive electronic devices [8].

b. *Energy efficiency*: Efficient operation of electrical equipment often requires a stable voltage supply at the rated frequency and waveform. Deviations from these parameters can result in increased energy consumption, reduced efficiency, and unnecessary heat losses in equipment.

c. *Economic impact*: Power quality issues can lead to production disruptions, equipment failures, and increased maintenance costs for industries. In commercial settings, poor power quality can impact transaction processing, data centers, and communication networks. These disruptions can result in significant financial losses [9].

d. *Voltage stability and balance*: Fluctuations in voltage levels can impact the stability of the power system and the voltage supplied to consumers. Unstable voltage levels can lead to equipment malfunction, flickering lights, and decreased equipment lifespan.

e. *Harmonics and power factor*: The presence of harmonics (nonlinear voltage and current components) and poor power factor (a measure of how effectively power is utilized) can lead to increased losses, lower system efficiency, and decreased system capacity.

f. *Electromagnetic interference (EMI)*: Poor power quality can lead to electromagnetic interference, which can affect the proper operation of electronic devices, communication systems, and other sensitive equipment [10].

g. *Voltage profile and voltage regulation*: Maintaining a stable voltage profile throughout the distribution network is essential to ensure the proper functioning of equipment and minimize voltage-related issues, such as overvoltage or undervoltage conditions.

h. *Renewable energy integration*: With the increasing integration of renewable energy sources like solar and wind, power quality becomes more critical. Fluctuating output from these sources can introduce voltage and frequency variations, requiring grid management strategies to maintain overall power quality.

i. *Customer satisfaction*: Consistently providing reliable power with good quality is essential for customer satisfaction. Homeowners, businesses, and industries rely on uninterrupted power supply and consistent performance of their electrical equipment.

j. *Regulatory compliance*: Power utilities often have to adhere to regulations and standards that define acceptable levels of power quality. Failure to meet these standards can result in penalties and regulatory actions.

k. *Public safety*: Poor power quality can pose safety risks, especially when it affects critical systems like emergency lighting, fire alarm systems, and medical equipment.

According to the mentioned things, it can be said that the purpose of writing this chapter of the book is to investigate the role of electric vehicles in improving power quality and how it affects them. The organization of this chapter is as follows. After the introduction in Section I, Section II deals with the concept and definition of power quality in the power system. In Section III, electric vehicles and their effects on important parts of power quality, including frequency and voltage, will be discussed. At the end and in Section IV, the findings will be categorized and summarized.

2. Power quality problem

The concept of 'power quality' pertains to ensuring that the power system maintains voltage and current waveforms that are nearly sinusoidal, with magnitudes and frequencies in line with specifications [11]. Power quality can be influenced by multiple factors, including deviations in voltage and frequency, imbalances, interruptions, flickering, and harmonics [12, 13]. Consequently, maintaining power quality holds significance in ensuring the dependable and secure functioning of smart grids, and this aspect is expected to be affected as the demand for EVs experiences imminent growth. The passage suggests that the increasing adoption of EVs could potentially impact power quality. The charging of EVs and their connection to the grid might introduce additional demands on the system, potentially leading to voltage and frequency variations or other PQ issues. Therefore, it is of great significance to investigate the PQ issues in smart grids while considering EVs [14–16]. Power quality issues in power systems can impact the reliability, efficiency, and overall performance of the electrical supply. The main power quality issues that power systems may encounter are shown in **Figure 2**.

- *Voltage sags and dips*: Also known as voltage drops, these are short-term reductions in voltage levels. They can be caused by faults, sudden changes in load, or other disturbances in the system. Voltage sags and dips can affect sensitive equipment and cause malfunctions or disruptions [17]. *Voltage swells*: Voltage swells are short-term increases in voltage levels. They can be caused by sudden load reduction, switching operations, or other disturbances. Like sags, voltage swells can stress equipment and lead to malfunction or damage.

- *Voltage fluctuations and flicker*: Rapid variations in voltage levels can result in flickering lights. These fluctuations are often caused by changing loads, such as the starting of large motors or other dynamic equipment.

- *Harmonic distortion*: Non-sinusoidal waveforms can introduce harmonics into the system. Harmonics are integer multiples of the fundamental frequency (50 or 60 Hz) and are often produced by nonlinear loads like computers, variable frequency drives, and other electronic devices. Harmonics can cause overheating in equipment and interfere with other equipment operations [18].

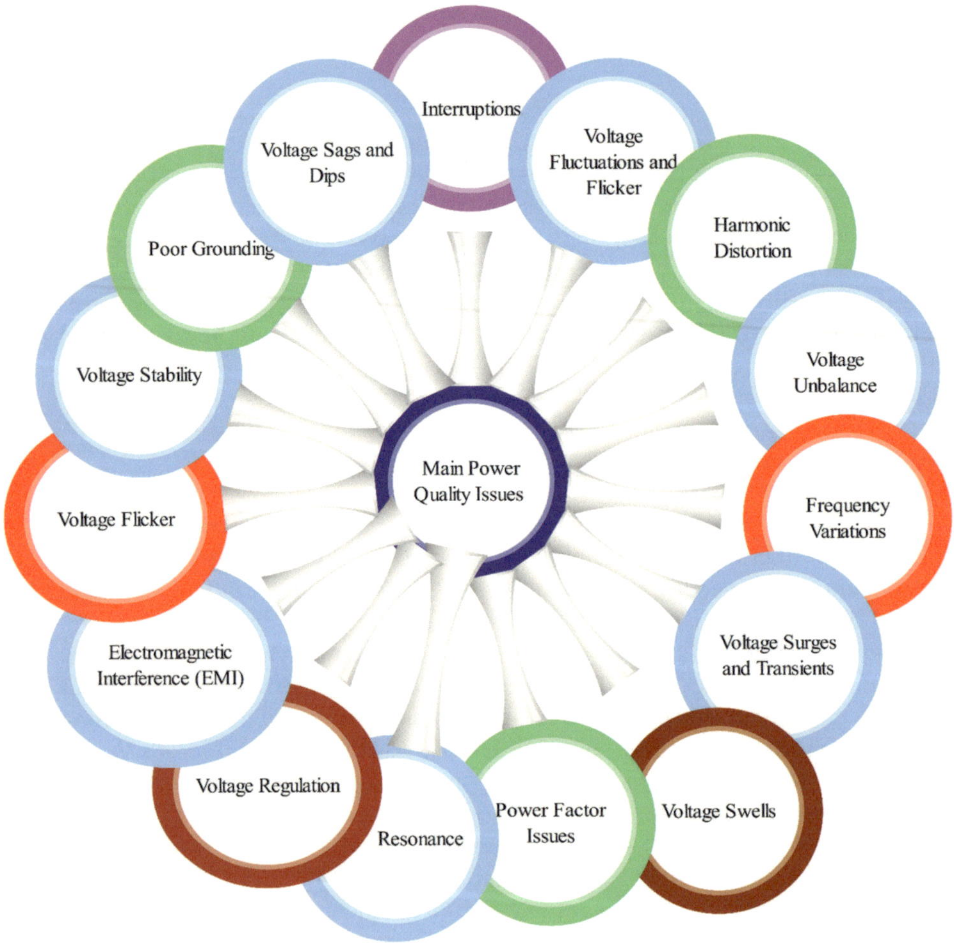

Figure 2.
The main power quality issues.

- *Voltage unbalance*: In three-phase systems, voltage unbalance occurs when there are significant differences in voltage levels among the three phases. This can lead to unequal distribution of power and increased losses.

- *Frequency variations*: Deviations from the standard frequency (50 or 60 Hz) can affect the operation of clocks, timers, and equipment that relies on precise timing.

- *Voltage surges and transients*: These are sudden and brief increases in voltage levels. They can be caused by lightning strikes, switching operations, or other disturbances. Voltage surges and transients can damage sensitive electronics and equipment.

- *Interruptions:* Power interruptions are complete losses of power supply and can be caused by faults, equipment failures, or maintenance activities. They can disrupt operations and cause inconvenience.

- *Power factor issues*: A low power factor indicates inefficient use of electrical power. It can result from reactive power consumption by devices like motors, leading to higher energy consumption and losses.

- *Voltage regulation*: Poor voltage regulation can lead to voltage levels that are outside the acceptable range, affecting equipment performance and potentially causing damage.

- *Resonance*: Resonance occurs when the natural frequency of a system matches the frequency of a disturbance, leading to excessive voltages or currents. Resonance can cause equipment failures.

- *Electromagnetic interference (EMI)*: EMI involves the generation of electromagnetic signals that can interfere with the proper functioning of sensitive electronic devices.

- *Voltage flicker*: Similar to fluctuations, voltage flicker is a repetitive variation in voltage levels, often caused by dynamic loads like arc furnaces.

- *Voltage stability*: Voltage stability issues can arise when the system is unable to maintain steady voltage levels, particularly during stressful conditions or extreme events.

- *Poor grounding*: Inadequate grounding can result in safety hazards, as well as introduce unwanted noise and distortions into the system [19].

Addressing these power quality issues requires a combination of proper design, equipment selection, maintenance practices, and monitoring tools to ensure the smooth and reliable operation of power systems. Pertaining to transient-based power quality (PQ) concerns, their classification can be based on their highest magnitudes, frequency components, as well as their characteristics concerning rise time and duration. Issues are categorized into short-duration and long-duration types based on their time spans and are further differentiated by the extent of magnitude fluctuations. The assessment of power system waveform distortion is often led by the analysis of total harmonic distortion (THD) and the harmonic spectrum. In terms of technical evaluation, the concept of intermittency serves as a valuable indicator for effectively identifying voltage fluctuations [20].

2.1 PQ issues in conventional power systems

Typically observed in conventional power systems characterized by sizeable centralized generators, overhead transmission and distribution lines, and passive user consumption, the primary instigators of power quality (PQ) issues are the presence of nonlinearities within loads or predominant system equipment. The ramifications of PQ problems encompass a range of consequences, including motor overheating, transformer saturation, system resonance, capacitor overloading, impaired protection mechanisms, fluctuations in light intensity, mechanical harm to generators and turbine shafts, production losses, and potential human health effects such as discomfort, irritation, and headaches [20].

2.2 PQ issues in modern power systems

Over recent decades, there has been a noticeable and continuous rise in the extensive deployment of distributed energy resources (DERs). Alongside this trend, there has been an increasing inclination toward employing power electronic-based interfaces. Unlike conventional power systems, where power electronic devices, acting as rigid nonlinear loads, have been used, their adoption has been more limited compared to contemporary power systems. The global drive to reduce emissions aligns with the growing adoption of electric vehicles (EVs), augmenting the significance of electrified transportation. Consequently, a transformation in both generation technology and the nature of electrical loads has occurred. Modern power systems are now subject to increased nonlinearity from both the generation and demand perspectives. The proliferation of electronic smart home appliances further accentuates concerns for power quality (PQ) issues among system operators.

Emerging factors such as adjustable speed drives (ASDs) and LED lamps introduce new elements that can adversely influence PQ metrics. The evolution of high-voltage direct current (HVDC) lines, the expansion of power line communications, and the ongoing shift from overhead lines to underground cables contribute to altering the traditional power system landscape. Additionally, the economic ramifications amplify the urgency for research that delves into the characteristics, origins, and techno-economic as well as environmental consequences of PQ concerns [20].

2.3 PQ problems associated with EV charging infrastructure

An electric vehicle (EV) charging station, also known as electric vehicle supply equipment (EVSE), serves the purpose of delivering electrical power to recharge the batteries of plug-in electric vehicles (PEVs) and other EVs. These charging stations are categorized based on the characteristics of the charging current, the speed of charging, and their geographical placement. A typical public EV charging station encompasses components such as a grid power source, a dedicated transformer, power quality measurement instruments, switchgear panels, chargers, and more. Compliance with IEC 61000-4-30 necessitates the installation of power quality meters at the point of common coupling, equipped to assess parameters like harmonics (including supra harmonics), voltage fluctuations such as sag and swell, as well as voltage flicker and disruptions [21]. Charging stations can be differentiated into AC charging and DC charging, contingent on the location of the AC/DC conversion process. AC charging occurs when the AC/DC conversion transpires within the vehicle itself. Conversely, if the AC/DC conversion occurs externally, within the EVSE, it is categorized as DC charging.

Charging stations can be categorized based on the charging speed or the duration required to attain a 100% state of charge (SoC) for the battery. These classifications include slow charging and fast charging. Slow charging refers to the method of charging that typically occurs overnight, spanning approximately 6–8 hours or more to fully charge an EV battery pack. In contrast, fast charging denotes a setup that expedites the charging process, requiring considerably less time. Fast charging setups are characterized by high-power output, enabling a vehicle to achieve an 80% charge within approximately 15 minutes. Currently, such charging stations are prevalent and typically possess power ratings within the range of 10–50 kW [22]. Charging solutions operating at the 100 kW level may become available in the medium-term timeframe. The rate of charging is contingent upon both the power rating of the EVSE and the capacity of the battery. A higher power rating results in a swifter charging rate. When the charging time is less than 30 minutes, the charger is commonly referred to as an ultrafast, superfast, flash charging, or rapid charger.

Charging stations can be categorized based on the location of the charging process, differentiating between home/residential charging and public charging stations.

Home charging is a prevalent practice among EV owners due to its cost-effectiveness. This approach generally employs slow chargers, taking advantage of the flexibility in available charging time. EV owners often opt to charge their vehicles overnight at their home EVSE, capitalizing on lower electricity tariffs compared to the commercial or industrial rates applicable in public charging infrastructure. Approximately 80% of EV charging occurs within the confines of the owner's home charging station.

Public charging stations, on the other hand, are facilities established and operated by utility companies or private entities with relevant licenses. These stations provide charging services for EVs beyond residential settings. The concept of vehicle to grid (V2G) involves EVs supplying power back to the grid during peak demand periods, effectively functioning as prosumers. Prosumers are entities that both consume electric power from the grid and contribute power back to the grid at the same point of connection (PoC) [21]. This practice of V2G has the potential to curtail peak power generation requirements and foster a balance between electricity demand and generation.

3. EVs for PQ improvement

Before we discuss the impact of EVs on power quality, it is necessary to know the different types of EVs. Nowadays, we can encounter different types of EVs, according to their engine technology. In general, they are sorted into five types as shown in **Figure 3**.

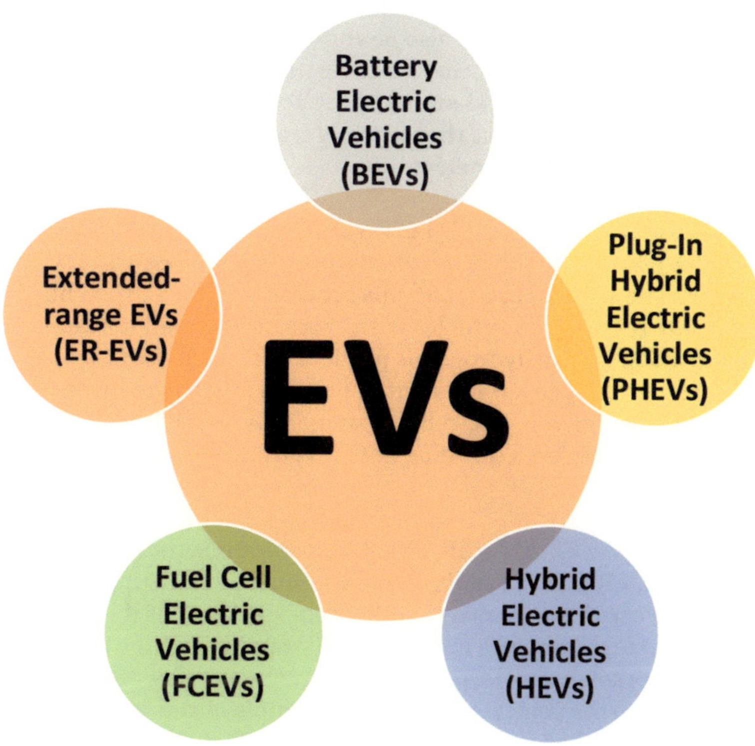

Figure 3.
EVs taxonomy.

- *Battery EVs (BEVs):* BEVs exclusively rely on electric power for their propulsion. Unlike internal combustion engine vehicles, BEVs lack an internal combustion engine and do not utilize liquid fuels of any kind [23]. Typically, BEVs utilize sizeable battery packs to ensure a satisfactory driving range. A standard BEV can cover distances ranging from 160 to 250 kilometers on a single charge, although certain models can achieve up to 500 kilometers with a single charge. A notable illustration of this vehicle category is the Nissan Leaf [24], a fully electric vehicle equipped with a 62 kWh battery that provides a driving range of 360 kilometers.

- *Plug-in hybrid electric vehicles (PHEVs):* PHEVs operate on a hybrid propulsion system consisting of a conventional internal combustion engine and an electric motor. The electric motor can be charged from an external electric source *via* a plug connection. PHEVs are engineered to store a sufficient amount of grid-supplied electricity, thereby substantially reducing their reliance on conventional fuels during typical driving scenarios. An instance of this category is the Mitsubishi Outlander PHEV [25], which is furnished with a 12 kWh battery enabling an electric-only range of approximately 50 kilometers. It's important to note, however, that the fuel efficiency of PHEVs often surpasses the figures provided by manufacturers [26].

- *Hybrid electric vehicles (HEVs):* HEVs operate through a dual power system, combining a conventional internal combustion engine with an electric motor. Unlike plug-in hybrid electric vehicles (PHEVs), HEVs lack the capability to be connected to the electrical grid for recharging. Notably, the energy source for the electric motor in HEVs is the vehicle's combustion engine. In more contemporary models, the batteries can also be charged through regenerative braking, where kinetic energy is transformed into electrical energy. A notable example is the fourth-generation hybrid model of the Toyota Prius [27], which features a 1.3 kWh battery. This battery theoretically grants the vehicle an electric-only driving range of up to 25 kilometers.

- *Fuel cell electric vehicles (FCEVs):* FCEVs are equipped with an electric motor that operates using a combination of compressed hydrogen and oxygen drawn from the air. The sole byproduct of this process is water [28]. While FCEVs are commonly regarded as vehicles with "zero emissions," it is important to note that although some hydrogen is produced from environmentally friendly sources (green hydrogen), a substantial portion is extracted from natural gas. An illustrative instance of this vehicle category is the Hyundai Nexo FCEV [29], capable of covering a distance of 650 kilometers without the need for refueling.

- *Extended-range EVs (ER-EVs):* ER-EVs closely resemble those in the BEV category. However, ER-EVs are additionally equipped with a supplementary combustion engine capable of recharging the vehicle's batteries if required. It is worth noting that unlike the engines found in PHEVs and HEVs, the combustion engine in ER-EVs serves solely for the purpose of battery charging and is not mechanically connected to the vehicle's wheels. An example of this vehicle type is the BMW i3 [30], furnished with a 42.2 kWh battery that allows an electric driving range of 260 kilometers. Additionally, users can access an extended-range mode, granting an extra 130 kilometers of travel distance.

Electric Vehicles (EVs) have the potential to enhance power quality in power systems through their integration with smart charging and vehicle-to-grid (V2G) technologies. By leveraging EV batteries as distributed energy storage resources, power systems can benefit from load balancing, voltage regulation, and frequency stability. These capabilities enable EVs to absorb excess power during periods of low demand and inject it back into the grid during peak hours, effectively reducing grid stress and enhancing overall stability. Additionally, EVs can participate in demand response programs, allowing their charging rates to be adjusted based on grid conditions. This assists utilities in managing peak loads and mitigating voltage fluctuations. The integration of EVs also incentivizes grid modernization efforts, fostering the deployment of advanced communication systems and grid management tools that further improve power quality. Overall, as EV adoption continues to grow, their intelligent interaction with the grid presents an opportunity to enhance power quality and reliability while supporting the integration of renewable energy sources. In this section, we will discuss the impact of EVs and the main components of power quality, i.e., voltage and frequency.

3.1 Role of EVs in voltage section

The dynamic charging behavior of electric vehicles (EVs) introduces a substantial stability challenge to power system networks. The substantial demand for EV charging places significant stress on utilities worldwide. For instance, in North America, power companies had to construct additional generating facilities to accommodate the augmented load from EVs [31]. The proliferation of EVs poses a notable challenge to voltage stability, particularly when they achieve high market penetration [32]. Previous research endeavors have consistently highlighted that EV charging exacerbates voltage instability. This issue is exacerbated by the presence of single-phase loads from EVs within distribution networks, leading to voltage imbalances. Moreover, the substantial inrush current triggered by EV charging gives rise to voltage dips for residential consumers, frequently coinciding with peak household load periods [33]. Investigations concerning the European Union's low voltage network unveiled that controlled charging often led to voltage violations at the feeder endpoints [34].

In Ref. [35], researchers delved into the voltage imbalance issue stemming from suboptimal charging methods. The investigation explored two distinct charging approaches: uncontrolled charging and tariff-based charging. The findings underscored that uncontrolled charging results in voltage imbalances surpassing a critical threshold of 2%. To address this, a tariff-based charging strategy was implemented, aiming to incentivize EV owners to charge during off-peak hours when electricity tariffs are lower. This approach was primarily designed to alleviate congestion during peak periods. The introduction of a substantial number of EVs has led to voltage drops and power losses in distribution, exacerbated by traffic congestion on road networks [36].

An examination of the charging patterns of plug-in hybrid EVs (PHEVs) was conducted to assess their influence on load profiles due to EV integration. The studies indicated that charging at public locations led to a 24–29% rise in daily energy consumption compared to home charging [37]. An evaluation of the Canadian distribution system concluded that the impact of fast charging systems is more pronounced than that of slow chargers [38]. Research on the Chinese distribution system highlighted that uncoordinated charging of EVs at different penetration levels resulted in voltage stability issues and phase imbalances. A parallel study adopted a smart charging approach to optimize charging cycles and ensure uninterrupted power flow within the system [39]. In Ref. [40], the analysis demonstrated that when EV penetration reached 0.28%, certain distribution networks encountered both low voltage stability concerns and transformer overload breaches. Although the present level of EV penetration did not forecast any issues, escalating penetration levels brought forth both low voltage instability and overload infractions for varying adoption rates. Particularly, at an EV penetration of 37.68%, the network faced complete collapse due to inadequate voltage regulation. Simulation results prominently identified voltage instability as the primary contributor to overall instability.

Charging electric vehicles (EVs) can induce voltage-related issues in the power network, encompassing voltage drops and fluctuations. A multitude of scholars have undertaken investigations into the repercussions of EV charging on network voltage. To account for the uncertainties surrounding EV availability, the Monte Carlo simulation method was employed in one study to scrutinize how EV charging impacts system voltage [41]. The extent of these impacts can either be substantial or negligible, contingent on factors such as the quantity of EVs, network attributes, and the characteristics of EV charging [42].

Research endeavors have yielded divergent outcomes. For instance, a study [43] underscored that the effect of the charging power of these vehicles on an urban distribution network remains minor when EV penetration is low. Conversely, in contrast to the findings in Ref. [43], another investigation [39] established that with a 50% EV penetration rate, network voltage deviations surpass established standards. Furthermore, in Ref. [44], outcomes revealed that the deployment of six EVs equipped with fast charging technology leads to violations of the prescribed network voltage standard.

Various research studies have explored the utilization of voltage regulation techniques and strategies to uphold network voltage within specified norms. Traditional methodologies, such as the strategic placement of optimal capacitor banks and the implementation of tap changers, can be employed to counteract voltage deviations within the network [45]. An alternative avenue involves managing the charging of electric vehicles (EVs) to enhance network voltage profiles. In this context, Masoum et al. [46] investigated a sophisticated control strategy for load management, focusing on the coordinated charging of EVs. The findings of this study highlighted that the synchronization of EV charging, facilitated by a practical tariff structure and a prioritized time zone scheme, effectively contributes to the voltage regulation of the system.

Furthermore, an effective strategy for managing electric vehicle (EV) chargers involves the decoupled control of active and reactive power, which proves instrumental in network voltage regulation. The active power control mechanism facilitates the manipulation of EV charging operations, while the reactive power control facet contributes by infusing reactive power into the grid to fortify network voltage levels. In Ref. [47], a prototype of a three-phase off-board EV charger was devised and put into operation. This charger not only maintains a steady DC-link voltage but also adeptly regulates network voltage within acceptable thresholds. This is achieved through the injection of reactive power, all the while ensuring no detrimental impact on the charging process. In the contemporary context, the management of EV charging to enhance voltage profiles stands as a viable consideration for both distribution and transmission service providers.

- *LV distribution networks*: Primary challenges include voltage drops during peak loads and voltage elevation caused by active power injection from photovoltaic (PV) systems. To address the voltage elevation issue, Marra et al. [48] introduced a coordinated approach between electric vehicle (EV) loads and PV systems, illustrated through a simulation involving a residence with rooftop PV and an EV. The simulation outcomes underscored the effectiveness of this approach [48]. Furthermore, Alam et al. [49] assessed this coordination strategy on an actual Australian distribution system employing genuine PV and EV data. Conversely, Akhtar et al. [50] proposed an alternative solution involving smart loads equipped with back-to-back converters to mitigate voltage disturbances arising from EVs and PVs in LV networks. The findings revealed that the application of smart loads alongside such converters could adeptly regulate bus voltage, albeit necessitating the use of two converters [50].

- *MV-level networks*: In the context of MV-level networks, García-Triviño et al. [51] conducted a study focused on voltage regulation employing a rapid charging station alongside a decentralized energy management system (EMS). The findings highlighted the viability of this innovative approach in voltage control through decentralized methods, independent of a communication interface [51]. Additionally, Torreglosa et al. [52] delved into a similar area by investigating a decentralized EMS for a charging station. This study employed model predictive control (MPC) to effectively regulate bus voltage within MV networks.

- *Transmission network:* An investigation [53] explored the potential contribution of end-user devices, such as electric vehicles (EVs), in the voltage control of the grid. The study introduced an active support group for frequency regulation and a reactive support group to oversee voltage control at the transmission bus. In a separate study, Rana et al. [54] explored a modified droop control technique incorporating EV aggregators, wind and solar units, and diesel generators to manage the frequency within a microgrid. The EV aggregator, in this scenario, adjusts the charging and discharging rates to align with set points while considering EVs and state of charge (SoC) limits [54]. Additionally, EV converters can inject active and reactive power into the grid to provide support during voltage ride-through conditions. During instances of renewable energy-related transients, EVs can mitigate transient stress on the grid by injecting active power. A subsequent study [55, 56] underscored the ability of EVs to effectively regulate both reactive and active power during voltage ride-through scenarios.

3.2 Voltage unbalance mitigation in electric distribution system with EVs

A phase reconfiguration strategy offers a means to mitigate voltage imbalance concerns. In Ref. [57], the financial implications of adopting a phase reconfiguration approach to alleviate the adverse effects of unbalanced electric vehicles (EVs) on a low voltage (LV) distribution system were examined. The results affirmed that employing a phased reconfiguration strategy, in conjunction with a time-of-use tariff, could effectively mitigate the imbalance impact of EVs. Furthermore, effective management of EV charging and discharging can contribute to alleviating phase unbalance issues [58]. The process of achieving voltage balance involves making optimal decisions regarding connection points (phases a, b, or c), charging and discharging power rates, as well as the charging and discharging status. In the pursuit of coordinated smart charging for all EVs and boosting the load profile tied to the electrical grid network, a control mechanism is under development [59]. This coordination relies on a communication infrastructure enabling data collection and transmission between EVs and an aggregator.

To standardize voltage profiles, the load profile is smoothed. The implementation of voltage regulators can effectively reduce network imbalance indices. Improved power quality (PQ) concerning voltage imbalances and variations can be attained through the use of energy storage devices, feeder capacitors, and D-STATCOM [60].

It's noteworthy that some EVs may occasionally experience elevated peak voltages. Unlike a central controller, a decentralized controller charges EVs locally, thereby minimizing communication infrastructure costs [61]. The electric vehicle current charger does not consider the neutral current stemming from phase voltage imbalances. Given the dynamic nature of input from distributed energy resources, both distributed energy resources and EVs necessitate coordination to address neutral current and voltage imbalance [62, 63]. An analysis of the merits and drawbacks of centralized and decentralized EV charging coordination models is presented. In the context of smart charging within the electrical distribution system, active power balance can be achieved through the implementation of the droop controller topology. Although power electronics devices like rectifiers and switching converters contribute reactive power into the distributed network, potentially burdening the distribution network, a droop controller-based mechanism was described in Ref. [64] to achieve a balanced system. Reactive power correction can be attained through auxiliary equipment specified in Ref. [65]. To mitigate voltage imbalance, techniques for generating zero and negative sequence voltages are employed. In **Table 1**, various approaches to mitigate the impact of EVs on voltage quality are summarized.

3.3 Role of EVs in frequency section

In power systems, maintaining a real-time equilibrium between generation and load is crucial to prevent deviations in grid frequency from the established norm. When a substantial charging load from electric vehicles (EVs) is introduced to the grid, additional power generation becomes imperative to uphold grid frequency

Impacts	Mitigation techniques	Description
Unbalanced voltage impact	Voltage regulators, phase reconfiguration technique, EV charging and discharging management.	Optimizing EV and phase reconfiguration techniques to mitigate voltage imbalances. Voltage unbalance can be reduced by utilizing flywheels, supercapacitors, battery storage systems, capacitive energy storage systems, superconducting magnetic storage systems, DVRs, and DSTATCOMs [66, 67].
Impact on voltage fluctuations	Control strategies for active and reactive power, voltage regulators, and charging management.	Voltage fluctuations can be reduced with transformer tap changes and capacitor banks. In order to reduce voltage fluctuations, EV charging is coordinated in a controlled manner to regulate network voltage [47, 66, 68].
Impacts on voltage profile	• Traditional voltage regulators. • Charging management of EVs. • Control strategy for active and reactive power that is decoupled.	Voltage regulation [46, 47]: Feeder capacitor bank. Tap changer. A smart load management control [57]: Coordinating EV charging. EV charging provides active power control, and reactive power injection regulates network voltage by injecting reactive energy into the grid [67, 69].

Table 1.
Approaches to mitigate impact of EVs on voltage quality.

within acceptable limits [70]. Moreover, the unpredictability inherent in the departure and arrival times of EVs introduces uncertainty into power systems, amplifying uncertainty from the demand side [71]. To rectify this, strategies such as load curtailment or demand-side management (DSM) initiatives can be implemented to align load and generation within the grid. Employing coordinated control of EVs serves to harmonize power distribution within the power system [70, 72]. Notably, EV aggregators possess the potential to play a significant role across energy and ancillary service markets. In terms of ramping capability, EVs outperform gas turbines, and they offer a more cost-effective alternative in comparison to traditional energy storage systems [70, 71, 73, 74].

As a subset of energy storage technology, electric vehicles (EVs) hold promising potential for active involvement in frequency regulation (FR) systems. This potential stems from the widespread proliferation of EVs and the continual advancements in vehicle-to-grid (V2G) technology. By adeptly adjusting their charging and discharging rates, EVs can effectively counteract frequency fluctuations originating from intermittent energy sources like wind and solar power. Research affirms that a smart grid can receive substantial frequency support through the utilization of EV fleets, harnessed as manageable loads. Employing a multiple model predictive control approach, Pahasa and Ngamroo [75] address EV charging/discharging and state-of-charge (SOC) control to stabilize microgrid frequency fluctuations. This methodology capitalizes on the expansive EV fleet, significantly enhancing the capabilities of V2G. Similarly, in Ref. [76], a virtual synchronous generator mechanism is integrated into an EV charging station, governing the bidirectional power flow to reinforce frequency stability. In Ref. [77], primary frequency control *via* V2G is accomplished by combining droop control and inertial response, facilitated by SOC management.

In the process of system frequency regulation (FR), electric vehicles (EVs) exhibit swift responsiveness to frequency deviations by adjusting their charging and discharging activities. However, it's imperative to acknowledge that while EVs can play a role in FR, their fundamental purpose as modes of transportation introduces considerations related to traffic patterns and drivers' demands. The manageable energy of EVs, influenced by driving regulations, demonstrates temporal variability within a day, thereby impacting the efficacy of system FR. Addressing this, Janjic and Velimirovic [78] establishes day-ahead optimal EV allocation for FR utilizing queuing theory and a fuzzy multi-criteria approach, yet this scheduling approach overlooks the inherent driving regulations of EVs. Given the essential transportation function of EVs, achieving driving demands remains essential in participating in optimal scheduling or FR endeavors. An optimal scheduling strategy aimed at EV charging and discharging is devised using mixed integer linear programming [79] to regulate system frequency; however, specific EV driving rules are omitted from consideration. In Ref. [80], an aggregator computes the FR capacity and anticipated FR output of EVs, categorizing their charging behaviors into maintaining or adjusting the battery state of charge (SOC). Nonetheless, the timing of EVs' grid connection is at odds with daily driving regulations.

Furthermore, certain existing research on frequency regulation (FR) strategies overlook the dynamic changes in electric vehicles' (EVs) manageable energy, resulting in impractical distribution of FR demand power. Load frequency control (LFC) models, as presented in Refs. [81, 82], involve EVs in secondary FR *via* distinct control approaches. However, in these literature instances, conventional units and EVs are engaged in FR independently, without considering their coordination. In a coordinated FR strategy involving both EVs and traditional units, the accurate calculation of EVs' controllable energy profoundly influences the effectiveness of FR demand power allocation. Many pertinent studies have either neglected or simplified factors such as traffic patterns and vehicle owners' demands, leading to inaccurate estimations of EVs' real-time manageable energy. Furthermore, due to the inherent variability in vehicle usage, EVs' controllable energy fluctuates dynamically. When the distribution ratio of FR demand power between EVs and traditional unit's remains fixed, the energy storage potential of EVs remains underutilized, and the FR impact becomes suboptimal.

Zhang et al. [83] conduct a comprehensive analysis of state transition behaviors within the context of the FR process. Building upon this foundation, the study formulates dynamic models for real-time controllable capacity and energy of electric vehicles (EVs). Subsequently, an adaptive scheme is introduced to dynamically adjust system FR coefficients in response to EVs' controllable energy, thereby enhancing the rationality of FR demand power allocation. Furthermore, a selection strategy is outlined for identifying specific EVs to partake in FR while accounting for driver preferences. The simulation outcomes affirm the viability and efficacy of the proposed methodologies.

The swift reactivity of electric vehicles (EVs) enhances the dynamic capabilities of the load frequency control (LFC) system. The engagement of EVs in frequency regulation services also opens avenues for supplementary economic gains within the frequency regulation markets. Consequently, the integration of EVs into frequency regulation services is poised for imminent encouragement. A novel concept, known as an "EV aggregator," introduced in Ref. [84], addresses the minimal regulation capacity required in frequency regulation markets. However, the participation of EV aggregators in frequency regulation services can introduce time-varying delays to the LFC problem. Researchers in Ref. [85] investigate delay-dependent stability within LFC incorporating EV aggregator participation in frequency regulation services. The stability analysis of time-delayed LFC systems with EV aggregators is tackled in Ref. [86], leveraging Lyapunov theory and the linear matrix inequality (LMI) technique. Moreover, the ramifications of load sharing between EV aggregators and conventional generators on stability delay margins are explored in [87, 88] within a multi-area power system context. In scenarios where a substantial number of EVs are aggregated within the EV aggregator, the latter must possess Automatic Generation Control (AGC) capabilities to adjust each EV's power output in real time. This entails a communication network for control signal exchange between the EV aggregator and EVs in the AGC system. Ref. [89] outlines a model for the AGC system comprising multiple EV aggregators, accounting for time-delay characteristics in response to control signals while considering the communication network within each area of the power system's EV aggregator domain. This approach not only enhances dynamic performance but also significantly reduces communication overhead and computation expenses, all while maintaining control efficiency.

The literature offers several approaches related to electric vehicles (EVs) engagement in power frequency control (PFC). For instance, Floch et al. [90] delve into the optimal EV charging to shape the load. In Ref. [91], a model-free controller design strategy for EVs is proposed. Sun et al. [92] present a control strategy relying on filters for EVs. A risk-constrained decision tool is put forth in Ref. [93] to incorporate EV aggregators into frequency regulation and energy markets. Latif et al. [94] suggest fractional order controllers for EVs' participation in secondary frequency control (SFC) within microgrids. Investigating interconnected power systems with thermal and hydropower plants featuring fuzzy fractional controllers, Arya [95] analyzes the influence of EVs on SFC. Arani and Mohamed [96] explore the cooperative provision of primary frequency control (PFC) by wind turbines and EVs. A state-space model facilitating real-time power control for aggregated EVs with minimal communication requirements is proposed in Ref. [97]. In Ref. [98], a stability margin-based strategy tunes the frequency-droop controllers of EVs for PFC participation. Moghadam et al. [99] investigate the randomization of EVs for PFC contribution. While numerous strategies are suggested for EV participation in PFC, their impact on generating unit wear and tear has yet to be explored.

4. Discussion and conclusion

In this chapter, the role of EVs in improving the voltage and frequency of power systems was analyzed. In conclusion, we can say that EVs have the potential to contribute to voltage stability in power systems by providing voltage regulation, reactive power support, and load management services. Strategic deployment and integration of EVs, along with appropriate technological and regulatory developments, can help enhance grid voltage stability and overall system reliability. Also, they have the potential to enhance frequency stability in power systems by providing demand response, energy storage, and ancillary services. However, their integration requires careful planning, technological development, and regulatory support to maximize their benefits while mitigating potential drawbacks. The classification of mechanisms affecting EVs in voltage stability is shown in **Figure 4**.

1. *Voltage regulation*: EVs with bidirectional charging capabilities, enabled by vehicle-to-grid (V2G) technology, can adjust their charging and discharging rates in response to grid voltage fluctuations. When the grid voltage is high, EVs can absorb excess electricity, thereby reducing the voltage. Conversely, when the grid voltage is low, EVs can inject stored energy back into the grid, helping to raise the voltage to appropriate levels.

2. *Reactive power injection*: EVs equipped with power electronics can provide reactive power support to the grid. Reactive power is essential for maintaining voltage levels within the desired range. EVs can inject or absorb reactive power as needed, helping to regulate voltage and improve grid stability.

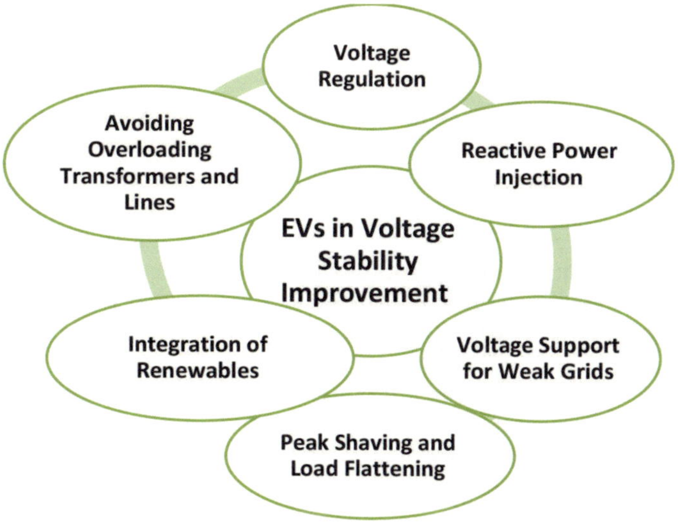

Figure 4.
Several mechanisms that EVs can contribute to voltage stability through them.

3. *Voltage support for weak grids*: In areas with weak distribution infrastructure, voltage levels might drop significantly during peak demand periods. EVs connected to these grids can provide local voltage support by discharging energy, thereby alleviating voltage drop issues and maintaining a stable supply.

4. *Peak shaving and load flattening*: Charging EVs during periods of low electricity demand and discharging them during peak demand periods can help flatten the load curve. This can contribute to maintaining more stable voltage profiles by reducing voltage variations associated with sudden load changes.

5. *Integration of renewables*: EVs can help integrate renewable energy sources by absorbing excess energy from intermittent sources during periods of high production and releasing it when renewable generation is low. This helps stabilize voltage levels as well.

6. *Avoiding overloading transformers and lines*: By using EVs as controllable loads, grid operators can manage the loading on transformers and distribution lines more effectively, preventing overloads and maintaining proper voltage distribution.

However, there are considerations to keep in mind:

- *Battery health:* Frequent cycling of EV batteries for voltage support purposes might impact battery health and longevity.

- *Coordination and control:* Effective coordination and control mechanisms are essential to ensure that a large number of EVs can provide voltage support in a harmonized manner without causing operational issues.

- *Standardization and infrastructure:* Like with any grid interaction technology, standardization of communication protocols and necessary charging infrastructure are critical for successful integration.

Electric vehicles (EVs) can potentially participate in Frequency Regulation (FR) strategies through three distinct approaches:

1. *Localized decision making*: The first method involves individual chargers configuring their charge/discharge power based on local information such as load variations, EV arrival schedules, and the local frequency of the area [100, 101].

2. *Decentralized FR signals*: In the second configuration, FR signals are generated according to operating voltage and power loss, and then transmitted to aggregators. These aggregators subsequently forward the signals to a control center in a decentralized manner [102, 103].

3. *Centralized decision mechanism*: The third approach commonly employed for FR involves a centralized decision mechanism. In this scenario, a control center manages the chargers by utilizing a communication system [104, 105].

Frequency fluctuations can lead to a range of problems, including equipment damage, blackouts, and overall instability of the power grid. Electric vehicles play a pivotal role in maintaining frequency stability, as depicted in **Figure 5**.

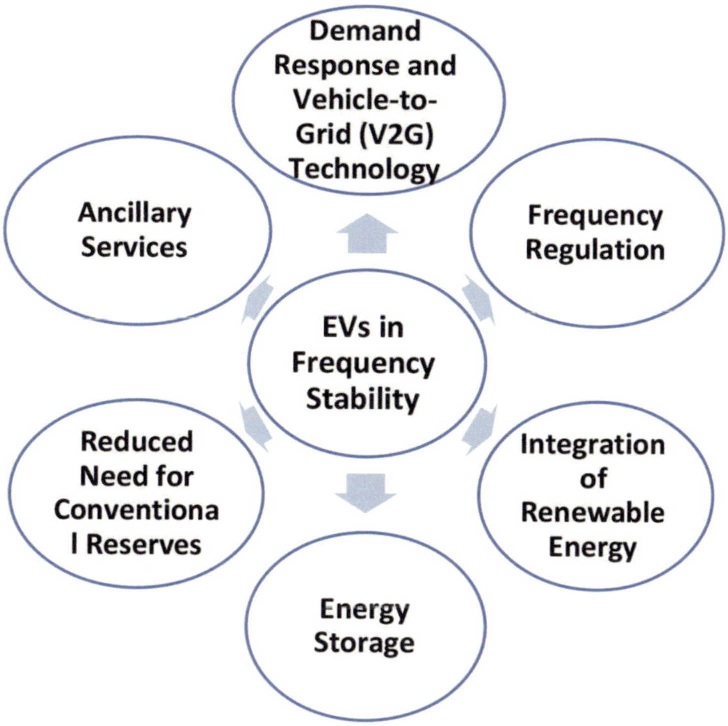

Figure 5.
Several mechanisms that EVs can contribute to frequency stability through them.

1. *Demand response and V2G technology*: EVs can act as a flexible load or even a source of power by utilizing V2G technology. During periods of high electricity demand, EVs can reduce their charging rate or temporarily discharge electricity back into the grid. This demand response capability helps balance supply and demand, stabilizing grid frequency.

2. *Frequency regulation*: Grid operators need to continuously match electricity generation with demand to maintain a stable frequency. EVs, when aggregated and controlled, can provide a distributed source of regulation services. They can respond rapidly to grid frequency changes by adjusting their charging or discharging rates, helping to stabilize frequency variations caused by sudden changes in generation or load.

3. *Integration of renewable energy*: Renewable energy sources like solar and wind are variable in nature. Their output depends on weather conditions, which can lead to fluctuations in the power supply. EVs can absorb excess electricity during periods of high renewable energy production and release it during low production periods, thereby helping to smooth out these fluctuations and maintain grid stability.

4. *Energy storage*: EVs effectively function as mobile energy storage units. When connected to the grid, their batteries can store excess electricity and inject it back into the grid when needed. This can mitigate frequency deviations caused by sudden changes in supply or demand.

5. *Reduced need for conventional reserves*: Traditional power plants often maintain spinning reserves to respond to sudden changes in demand. With a larger fleet of EVs equipped with V2G capabilities, these reserves could potentially be reduced as EVs can provide a distributed and responsive source of power.

6. *Ancillary services*: EVs can participate in providing ancillary services such as frequency regulation, voltage control, and reactive power support. These services are essential for maintaining stable grid operations.

However, there are also challenges to consider:

- *Battery degradation:* Frequent charging and discharging of EV batteries for grid services could accelerate their degradation, impacting the overall lifespan of the batteries.

- *Infrastructure and standardization:* Widespread implementation of V2G technology requires standardized communication protocols, infrastructure investment, and regulatory frameworks to ensure safe and efficient operation.

- *User preferences:* The ability to use an EV's battery for grid services might be subject to user preferences and priorities. Some EV owners might prioritize their vehicle's own battery life over grid services.

Author details

Mohammad Mehdi Amiri[1*], Saleh Aghajan-Eshkevari[1], Mohammad Ali Rahimi[1] and Ali Samari[2]

1 Faculty of Electrical Engineering, Shahid Beheshti University, Tehran, Iran

2 Faculty of Electrical Engineering, Bu-Ali Sina University, Hamadan, Iran

*Address all correspondence to: mehdi_hmf@yahoo.com

IntechOpen

© 2024 The Author(s). Licensee IntechOpen. This chapter is distributed under the terms of the Creative Commons Attribution License (http://creativecommons.org/licenses/by/3.0), which permits unrestricted use, distribution, and reproduction in any medium, provided the original work is properly cited.

References

[1] Ram G, Kardolus M. Roadmap Electric Vehicles and Grid Integration (V1G Versus V2G). Power Research Electronics BV; 2020. Available from: http://www.pr-electronics.nl/en/news/85/roadmap-electric-vehicles-and-grid-integration-v1g-versus-v2g/

[2] Haider S, Schegner P. Simulating the impacts of uncontrolled electric vehicle charging in low voltage grids. Energies. 2021;**14**(8):2330

[3] Jones CB, Lave M, Vining W, Garcia BM. Uncontrolled electric vehicle charging impacts on distribution electric power systems with primarily residential, commercial or industrial loads. Energies. 2021;**14**(6):1688

[4] Cheng Y, Ye S, Wang R, Gao H, Liu J. Joint planning of electric vehicle charging station and energy storage system in the distribution network. In: 2023 Panda Forum on Power and Energy (PandaFPE). IEEE; 2023. pp. 225-230. Available from: https://ieeexplore.ieee.org/abstract/document/10140862

[5] Jawad S, Liu J. Electrical vehicle charging load mobility analysis based on spatial-temporal model in traffic-distribution networks. In: 2023 Panda Forum on Power and Energy (PandaFPE). IEEE; 2023. pp. 700-704. Available from: https://ieeexplore.ieee.org/abstract/document/10140563

[6] Shah TH, Shabbir A, Waqas A, Janjua AK, Shahzad N, Pervaiz H, et al. Techno-economic appraisal of electric vehicle charging stations integrated with on-grid photovoltaics on existing fuel stations: A multicity study framework. Renewable Energy. 2023;**209**:133-144

[7] Kippke MA, Arboleya P, El Sayed I. Advanced metering infrastructure for smart grid real-time energy management using mesh networks based in IEEE802.15.4 and 6LoWPAN. In: 2021 IEEE Madrid PowerTech. IEEE; 2021. pp. 1-6. Available from: https://ieeexplore.ieee.org/abstract/document/9494778

[8] Khan S, Singh B, Makhija P. A review on power quality problems and its improvement techniques. In: 2017 Innovations in Power and Advanced Computing Technologies (i-PACT). IEEE; 2017. pp. 1-7. Available from: https://ieeexplore.ieee.org/abstract/document/8244882

[9] Bagdadee AH, Zhang L. Power quality impact on the industrial sector: A case study of Bangladesh. Journal of Electrical Engineering & Technology. 2019;**14**:1841-1857

[10] Xu D, Li S, Li C, Gao Z. Active electromagnetic interference (EMI) filter for conducted noise reduction. In: 2022 IEEE 9th International Conference on Power Electronics Systems and Applications (PESA). IEEE; 2022. pp. 1-5. DOI: 10.1109/PESA55501.2022.10038337

[11] Chattopadhyay S, Mitra M, Sengupta S. Electric power quality. In: Electric Power Quality. Power Systems. Dordrecht: Springer; 2011. DOI: 10.1007/978-94-007-0635-4_2

[12] Kappagantu R, Daniel SA, Yadav A. Power quality analysis of smart grid pilot project, Puducherry. Procedia Technology. 2015;**21**:560-568

[13] Niitsoo J, Taklaja P, Palu I, et al. Power quality issues concerning photovoltaic generation and electrical

vehicle loads in distribution grids. Smart Grid and Renewable Energy. 2015;**6**:164

[14] Gandoman FH, Ahmadi A, Sharaf AM, et al. Review of FACTS technologies and applications for power quality in smart grids with renewable energy systems. Renewable and Sustainable Energy Reviews. 2018;**82**:502-514

[15] Anjana KR, Shaji RS. A review on the features and technologies for energy efficiency of smart grid. International Journal of Energy Research. 2018;**42**(3):936-952

[16] Monteiro V, Carmo JP, Pinto JG, et al. A flexible infrastructure for dynamic power control of electric vehicle battery chargers. IEEE Transactions on Vehicular Technology. 2016;**65**(6):4535-4547

[17] Boričić A, Torres JLR, Popov M. Fundamental study on the influence of dynamic load and distributed energy resources on power system short-term voltage stability. International Journal of Electrical Power & Energy Systems. 2021;**31**. DOI: 10.1016/J.IJEPES.2021.107141. Available from: https://www.sciencedirect.com/science/article/pii/S014206152100380X

[18] Kalair AR, Stojcevski A, Seyedmahmoudian M, Abas N, Kalair A, Khan N, et al. Steady-state and time-varying harmonics in distribution system. In: Uncertainties in Modern Power Systems. Academic Press; 2021. pp. 485-539. DOI: 10.1016/B978-0-12-820491-7.00014-1

[19] Ehrenwerth J. Electrical and fire safety. In: Anesthesia Equipment. WB Saunders. Elsevier; 2021. pp. 526-558

[20] Ahmadi A, Tavakoli A, Jamborsalamati P, Rezaei N, Miveh MR, Gandoman FH, et al. Power quality improvement in smart grids using electric vehicles: A review. IET Electrical Systems in Transportation. 2019;**9**(2):53-64

[21] Central Electricity Authority (Technical Standards for Connectivity of the Distributed Generation Resources) Amendment Regulations. 2019. Available from: https://cea.nic.in/regulations-category/connectivity-of-distributed-generation-resources/?lang=en

[22] Sivaraman P, Sharmeela C. IoT based battery management system for hybrid electric vehicle. In: Chitra A, Sanjeevikumar P, Bo J, Holm-Nielsen, Himavathi S, editors. Artificial Intelligent Techniques for Electric and Hybrid Electric Vehicles. WileyScrivener Publication; 2020. Available from: https://onlinelibrary.wiley.com/doi/abs/10.1002/9781119682035.ch1

[23] Ahmed AA, Nazzal MA, Darras BM, Deiab I. A comprehensive sustainability assessment of battery electric vehicles, fuel cell electric vehicles, and internal combustion engine vehicles through a comparative circular economy assessment approach. Sustainability. 2022;**15**(1):171

[24] insideEVs. Nissan Reveals LEAF e-Plus: 62 kWh Battery, 226-Mile Range. 2019. Available from: https://insideevs.com/nissan-reveals-leaf-e-plus-ces/ [Accessed: February 17, 2021]

[25] Mitsubishi Motors. Mitsubishi Outlander PHEV 2018. 2019. Available from: https://www.mitsubishicars.com/outlander-phev/2018/specifications [Accessed: February 17, 2021]

[26] Plötz P, Moll C, Bieker G, Mock P, Li Y. Real-World Usage of Plug-In Hybrid Electric Vehicles: Fuel Consumption,

Electric Driving, and CO2 Emissions; Technical Report. Washington, DC, USA: International Council on Clean Transportation Europe (ICCT); 2020

[27] The Car Guide. 2014 Toyota Prius PHV: To Plug In or Not to Plug In? 2014. Available from: https://www.guideautoweb.com/en/articles/21152/2014-toyota-prius-phv-to-plug-in-or-not-to-plug-in/ [Accessed: February 21, 2021]

[28] Endiz MS. A comparison of battery and hydrogen fuel cell electric vehicles for clean transportation. Orclever Proceedings of Research and Development. 2023;**2**(1):10-17

[29] Hyundai. All-New Hyundai NEXO—Technical Specifications. 2019. Available from: https://www.hyundai.news/eu/press-kits/all-new-hyundai-nexo-technical-specifications/ [Accessed: February 21, 2021]

[30] insideEVs. 2019 BMW i3, i3 REx, i3s & i3s REx: Full Specs. 2019. Available from: https://insideevs.com/2019-bmw-i3-rex-i3srex-full-spec/ [Accessed: February 21, 2021]

[31] Hadley SW. Impact of Plug-in Hybrid Vehicles on the Electric Grid. Oak Ridge, Tennessee: Oak Ridge National Laboratory; 2006. pp. 1-18

[32] Dharmakeerthi CH, Mithulananthan N, Saha TK. Impact of electric vehicle fast charging on power system voltage stability. International Journal of Electrical Power and Energy Systems. 2013;**57**:241-249

[33] Klayklueng T, Dechanupaprittha S. Impact analysis on voltage unbalance of EVs charging on a low voltage distribution system. In: Proceedings of the International Electrical Engineering Congress. Bangkok: IEEE; 2014

[34] Nour M et al. Impacts of plug-in electric vehicles charging on low voltage distribution network. In: International Conference on Innovative Trends in Computer Engineering (ITCE 2018). Egypt: Aswan University; 2018

[35] Haq AU et al. Impact of electric vehicle charging on voltage unbalance in an urban distribution network. Journal of Intelligent and Robotic System. 2015;**1**:51-60. Available from: https://link.springer.com/article/10.1007/s40903-015-0005-x

[36] Luo Y et al. The impact of large scale electric vehicles charging behaviour on distribution system and local traffic system. In: 7th IFAC Symposium on Advance in Automotive Control, the Internation Federation of Automotive Control. Tokyo, Japan: Elsevier; 2013

[37] Weiller C. Plug-in hybrid electric vehicle impacts on hourly electricity demanding the United States. Energy Policy. 2011;**39**(6):3766-3778

[38] Akhavan-Rezai E et al. Uncoordinated charging impacts of electric vehicles on electric distribution grids: Normal and fast charging comparison. In: Proceedings of IEEE Power and Energy Society General Meeting. San Diego, CA, USA: IEEE; 2012

[39] Li HL, Bai XM, Tan W. Impacts of plug-in hybrid electric vehicles charging on the distribution grid and smart charging. In: Proceedings of the IEEE POWERCON 2012: International Conference on Power System Technology. IEEE; 2012. Available from: https://ieeexplore.ieee.org/abstract/document/6401265

[40] Chophel K, Lhendup T, Chhetri R, Pradhan P. Impact of electric vehicle charging on low voltage network

stability. American Journal of Electrical and Electronic Engineering. 2020;**8**(4):131-137

[41] Amir M, Mateen S, Zaheeruddin, Haque A. Impacts assessment of ultra-fast EVs charging stations integrated with distribution energy system and mitigations measures using online controller. In 2023 International Conference on Recent Advances in Electrical, Electronics & Digital Healthcare Technologies (REEDCON). IEEE; 2023. pp. 403-408

[42] Qiu YL, Wang YD, Iseki H, Shen X, Xing B, Zhang H. Empirical grid impact of in-home electric vehicle charging differs from predictions. Resource and Energy Economics. 2022;**67**. DOI: 10.1016/J.RESENEECO.2021.101275. Available from: https://www.sciencedirect.com/science/article/abs/pii/S0928765521000609

[43] Rautiainen A, Mutanen A, Repo S, et al. Case studies on impacts of plug-in vehicle charging load on the planning of urban electricity distribution networks. In: 2013 Eighth International Conference and Exhibition on Ecological Vehicles and Renewable Energies (EVER). Monte Carlo, Monaco: IEEE; 2013. Available from: https://ieeexplore.ieee.org/abstract/document/6521524

[44] Yong JY, Ramachandaramurthy VK, Tan KM, et al. Bi-directional electric vehicle fast charging station with novel reactive power compensation for voltage regulation. International Journal of Electrical Power & Energy Systems. 2015;**64**:300-310

[45] Janiga K. A review of voltage control strategies for low-voltage networks with high penetration of distributed generation. Informatyka, Automatyka, Pomiary w Gospodarce i Ochronie Środowiska. 2020;**10**(3):60-65. DOI: 10.35784/IAPGOS.1928

[46] Masoum AS, Deilami S, Moses PS, et al. Smart load management of plug-in electric vehicles in distribution and residential networks with charging stations for peak shaving and loss minimisation considering voltage regulation. IET Generation Transmission and Distribution. 2011;**5**:877-888

[47] Yong JY, Fazeli SM, Ramachandaramurthy VK, et al. Design and development of a three-phase off-board electric vehicle charger prototype for power grid voltage regulation. Energy. 2017;**133**:128-141

[48] Marra F, Yang YG, Fawzy YT, et al. Improvement of local voltage in feeders with photovoltaic using electric vehicles. IEEE Transactions on Power Apparatus and Systems. 2013;**28**(3):3515-3516. Art. no. 6490073 (in English)

[49] Alam MJE, Muttaqi KM, Sutanto D. Effective utilization of available PEV battery capacity for mitigation of solar PV impact and grid support with integrated V2G functionality. IEEE Transactions on Smart Grid. 2016;**7**(3):1562-1571. Art. no. 7307203 (in English)

[50] Akhtar Z, Chaudhuri B, Hui SYR. Smart loads for voltage control in distribution networks. IEEE Transactions on Smart Grid. 2017;**8**(2):937-946 (in English)

[51] García-Triviño P, Torreglosa JP, Fernández-Ramírez LM, et al. Control and operation of power sources in a medium-voltage direct-current microgrid for an electric vehicle fast charging station with a photovoltaic and a battery energy storage system. Energy. 2016;**115**:38-48 (in English)

[52] Torreglosa JP, García-Triviño P, Fernández-Ramirez LM, et al. Decentralized energy management strategy based on predictive controllers for a medium voltage direct current photovoltaic electric vehicle charging station. Energy Conversion and Management. 2016;**108**:1-13 (in English)

[53] Bayat M, Sheshyekani K, Rezazadeh A. A unified framework for participation of responsive end-user devices in voltage and frequency control of the smart grid. IEEE Transactions on Power Apparatus and Systems. 2014;**30**:1369-1379

[54] Rana R, Singh M, Mishra S. Design of modified droop controller for frequency support in microgrid using fleet of electric vehicles. IEEE Transactions on Power Apparatus and Systems. 2017;**32**(5):3627-3636 (in English)

[55] Falahi M, Chou HM, Ehsani M, et al. Potential power quality benefits of electric vehicles. IEEE Transactions on Sustainable Energy. 2013;**4**(4):1016-1023. Art. no. 6541983 (in English)

[56] Kaur K, Rana R, Kumar N, et al. A colored petri net based frequency support scheme using fleet of electric vehicles in smart grid environment. IEEE Transactions on Power Systems. 2016;**31**(6):4638-4649

[57] Farahani HF. Improving voltage unbalance of low-voltage distribution networks using plug-in electric vehicles. Journal of Cleaner Production. 2017;**148**:336-346

[58] Verma A, Singh B. Multi-objective reconfigurable three-phase off-board charger for EV. IEEE Transactions on Industry Applications. 2019;**55**(4):4192-4203. DOI: 10.1109/TIA.2019.2908950

[59] Ahn C, Li CT, Peng H. Optimal decentralized charging control algorithm for electrified vehicles connected to smart grid. Journal of Power Sources. 2011;**196**(23):10369-10379. DOI: 10.1016/j.jpowsour.2011.06.093

[60] Kabir ME, Assi C, Alameddine H, Antoun J, Yan J. Demand-aware provisioning of electric vehicles fast charging infrastructure. IEEE Transactions on Vehicular Technology; 2020;**69**(7):6952-6963. DOI: 10.1109/TVT.2020.2993509

[61] García-Villalobos J, Zamora I, Knezović K, Marinelli M. Multi-objective optimization control of plug-in electric vehicles in low voltage distribution networks. Applied Energy. 2016;**180**:155-168

[62] Bozalakov DV, Laveyne J, Desmet J, Vandevelde L. Overvoltage and voltage unbalance mitigation in areas with high penetration of renewable energy resources by using the modified three-phase damping control strategy. Electric Power Systems Research. 2019;**168**:283-294. DOI: 10.1016/j.epsr.2018.12.001

[63] Kim S, Cha H, Kim H-G. High-efficiency voltage balancer having DC–DC converter function for EV charging station. IEEE Journal of Emerging and Selected Topics in Power Electronics. 2021;**9**(1):812-821. DOI: 10.1109/JESTPE.2019.2963124

[64] Knezović K, Marinelli M. Phase-wise enhanced voltage support from electric vehicles in a Danish lowvoltage distribution grid. Electric Power Systems Research. 2016;**140**:274-283. DOI: 10.1016/j.epsr.2016.06.015

[65] Leemput N, Geth F, Van Roy J, Büscher J, Driesen J. Reactive power support in residential LV distribution grids through electric vehicle charging.

Sustainable Energy, Grids and Networks. 2015;**3**:240-335. DOI: 10.1016/j.segan.2015.05.002

[66] Siti W, Nicolae DV, Abdul-Ganiyu J, Ukil A. Reconfiguration and load balancing in the LV and MV distribution networks for optimal performance. IEEE Transactions on Power Delivery. 2007;**22**:2534-2540. DOI: 10.1109/TPWRD.2007.905581

[67] Gray Matthew K, Morsi Walid G. Economic assessment of phase reconfiguration to mitigate the unbalance due to plug-in electric vehicles charging. Electric Power Systems Research. 2016;**140**:S0378779616302164. DOI: 10.1016/j. epsr.2016.06.008. Available from: https://www.sciencedirect.com/science/article/abs/pii/S0378779616302164

[68] Devi S, Geethanjali M. Optimal location and sizing determination of distributed generation and DSTATCOM using particle swarm optimization algorithm. International Journal of Electrical Power & Energy Systems. 2014;**62**:562-570. DOI: 10.1016/j.ijepes.2014.05.015

[69] Pillai JR, Bak-Jensen B. Impacts of electric vehicle loads on power distribution systems. In: 2010 IEEE Vehicle Power and Propulsion Conference. Lille, France: IEEE; 2010. pp. 1-6. Available from: https://ieeexplore.ieee.org/abstract/document/5729191

[70] Brooks A, Lu E, Reicher D, Spirakis C, Weihl B. Demand dispatch. IEEE Power and Energy Magazine. May-june 2010;**8**(3):20-29. DOI: 10.1109/MPE.2010.936349

[71] Tavakoli A, Negnevitsky M, Nguyen DT, et al. Energy exchange between electric vehicle load and wind generating utilities. IEEE Transactions on Power Apparatus and Systems. 2016;**31**:1248-1258

[72] Sortomme E, El-Sharkawi MA. Optimal scheduling of vehicle-to-grid energy and ancillary services. IEEE Transactions on Smart Grid. 2012;**3**:351-359

[73] Manbachi M, Farhangi H, Palizban A, et al. A novel volt-VAR optimization engine for smart distribution networks utilizing vehicle to grid dispatch. International Journal of Electrical Power & Energy Systems. 2016;**74**:238-251

[74] Rogers A, Henderson A, Wang X, et al. Compressed air energy storage: Thermodynamic and economic review. In: 2014 IEEE PES General Meeting–Conference & Exposition. National Harbor, MD, USA: IEEE; 2014. Available from: https://ieeexplore.ieee.org/abstract/document/6939098

[75] Pahasa J, Ngamroo I. PHEVs bidirectional charging/discharging and SOC control for microgrid frequency stabilization using multiple MPC. IEEE Transactions on Smart Grid. 2015;**6**(2):526-533

[76] Karanveer D, Mukesh S. Frequency support in a micro-grid using virtual synchronous generator based charging station. IET Renewable Power Generation. 2018;**12**(9):1034-1044

[77] Hernandez JC, Sanchez-Sutil F, Vidal PG, et al. Primary frequency control and dynamic grid support for vehicle-to-grid in transmission systems. International Journal of Electrical Power & Energy Systems. 2018;**100**:152-166

[78] Janjic A, Velimirovic LZ. Optimal scheduling of utility electric vehicle fleet offering ancillary services. ETRI Journal. 2015;**37**(2):215-246

[79] Kaur K, Kumar N, Singh M. Coordinated power control of electric vehicles for grid frequency support: MILP-based hierarchical control design. IEEE Transactions on Smart Grid. 2018;**10**(3):3364-3373

[80] Liu H, Hu Z, Song Y, et al. Vehicle-to-grid control for supplementary frequency regulation considering charging demands. IEEE Transactions on Power Systems. 2015;**30**(6):3110-3119

[81] Dutta A, Debbarma S. Frequency regulation in deregulated market using vehicle-to-grid services in residential distribution network. IEEE Systems Journal. 2018;**12**(3):2812-2820

[82] Khooban MH, Niknam T, Blaabjerg F, et al. A new load frequency control strategy for micro-grids with considering electrical vehicles. Electric Power Systems Research. 2017;**143**:585-598

[83] Zhang Q, Li Y, Li C, Li CY. Real-time adjustment of load frequency control based on controllable energy of electric vehicles. Transactions of the Institute of Measurement and Control. 2020;**42**(1):42-54

[84] Liu H, Huang K, Ni W, Qi J, Qiuwei W, Ma S, et al. Optimal dispatch for participation of electric vehicles in frequency regulation based on area control error and area regulation requirement. Applied Energy. 2019;**240**:46-55

[85] Ko KS, Sung DK. The effect of EV aggregators with time-varying delays on the stability of a load frequency control system. IEEE Transactions on Power Apparatus and Systems. 2018;**33**(1):669-680

[86] Zhou S, Zeng H, Xiao H. Load frequency stability analysis of time-delayed multi-area power systems with EV aggregators based on Bessel-Legendre inequality and model reconstruction technique. IEEE Access. 2020;**8**:99948-99955

[87] Naveed A, Sonmez S, Ayasun S. Impact of load sharing schemes on the stability delay margins computed by Rekasius substitution method in load frequency control system with electric vehicles aggregator. International Transactions on Electrical Energy Systems. 2021;**31**:e12884

[88] Naveed A, Sonmez S, Ayasun S. Impact of electric vehicles aggregators with communication delays on stability delay margins of two-area load frequency control system. Transactions of the Institute of Measurement and Control. 2021;**43**:2860-2871

[89] Tripathi S, Singh VP, Kishor N, Pandey AS. Load frequency control of power system considering electric vehicles' aggregator with communication delay. International Journal of Electrical Power & Energy Systems. 2023;**145**:108697

[90] Floch CL, Belletti F, Moura S. Optimal charging of electric vehicles for load shaping: A dual-splitting framework with explicit convergence bounds. IEEE Transactions on Transportation Electrification. 2016;**2**(2):190-199

[91] Liu Z, Pan H. Barrier function-based adaptive sliding mode control for application to vehicle suspensions. The IEEE Transaction on Transportation Electrification. 2021;7(3):2023-2033

[92] Sun W, Pan H, Gao H. Filter-based adaptive vibration control for active vehicle suspensions with electrohydraulic actuators. IEEE Transactions on Vehicular Technology. 2016;**65**(6):4619-4626

[93] Habibifar R, Aris Lekvan A, Ehsan M. A risk-constrained decision support tool for EV aggregators participating in energy and frequency regulation markets. Electric Power Systems Research. 2020;**185**, Art. no. 106367. Available from: https://www.sciencedirect.com/science/article/abs/pii/S0378779620301735

[94] Latif A, Das DC, Ranjan S, Barik AK. Comparative performance evaluation of WCA-optimised non-integer controller employed with WPG–DSPG–PHEV based isolated two-area interconnected microgrid system. IET Renewable Power Generation. 2019;**13**(5):725-736

[95] Arya Y. Effect of electric vehicles on load frequency control in interconnected thermal and hydrothermal power systems utilising CFFOIDF controller. IET Generation, Transmission & Distribution. 2020;**14**(14):2666-2675

[96] Arani MFM, Mohamed YA-RI. Cooperative control of wind power generator and electric vehicles for microgrid primary frequency regulation. IEEE Transactions on Smart Grid. 2018;**9**(6):5677-5686

[97] Wang M et al. State space model of aggregated electric vehicles for frequency regulation. IEEE Transactions on Smart Grid. 2020;**11**(2):981-994

[98] Izadkhast S, Garcia-Gonzalez P, Frias P, Bauer P. Design of plugin electric vehicle's frequency-droop controller for primary frequency control and performance assessment. IEEE Transactions on Power Apparatus and Systems. 2017;**32**(6):4241-4254

[99] Moghadam MRV, Zhang R, Ma RTB. Distributed frequency control via randomized response of electric vehicles in power grid. IEEE Transactions on Sustainable Energy. 2016;7(1):312-324

[100] Pal A, Bhattacharya A, Chakraborty AK. Placement of public fast-charging station and solar distributed generation with battery energy storage in distribution network considering uncertainties and traffic congestion. Journal of Energy Storage. 2021;**41**:102939

[101] Di Somma M, Graditi G, Siano P. Optimal bidding strategy for a DER aggregator in the day-ahead market in the presence of demand flexibility. IEEE Transactions on Industrial Electronics. 2018;**66**:1509-1519

[102] Nosratabadi SM, Hooshmand R-A, Gholipour E. A comprehensive review on microgrid and virtual power plant concepts employed for distributed energy resources scheduling in power systems. Renewable and Sustainable Energy Reviews. 2017;**67**:341-363

[103] Aurangzeb M, Xin A, Iqbal S, Jan MU. An evaluation of flux-coupling type SFCL placement in hybrid grid system based on power quality risk index. IEEE Access. 2020;**8**:98800-98809

[104] Iqbal S, Xin A, Jan MU, Abdelbaky MA, Rehman HU, Salman S, et al. Aggregation of EVs for primary frequency control of an industrial microgrid by implementing grid regulation & charger controller. IEEE Access. 2020;**8**:141977-141989

[105] Kong Q, Fowler M, Entchev E, Ribberink H, McCallum R. The role of charging infrastructure in electric vehicle implementation within smart grids. Energies. 2018;**11**:3362